Pittsburgh-Konstanz Series in the Philosophy
and History of Science

Theory and Method in the Neurosciences

EDITED BY Peter K. Machamer,
Rick Grush, and
Peter McLaughlin

University of Pittsburgh Press

Published by the University of Pittsburgh Press,
Pittsburgh, Pa. 15261
Copyright © 2001 University of Pittsburgh Press
All rights reserved
Manufactured in the United States of America
Printed on acid-free paper
10 9 8 7 6 5 4 3 2 1

ISBN 0-8229-4140-6

Contents

Theory and Method in the Neurosciences

Introduction

Peter K. Machamer, Rick Grush, and Peter McLaughlin

The 1990s were designated the Decade of the Brain. Surely the growth and development of brain research during that decade was remarkable, reminiscent of progress in molecular biology during the 1950s (Crick 1958). Medical research into drugs was certainly one of the more active areas of investigation. New medical diagnostic uses for such imaging techniques as positron emission tomography (PET), nuclear magnetic resonance imaging (nMRI), and functional magnetic resonance imaging (fMRI) were also developed, and these in turn provided sources of data for many other areas of research.

The field of cognitive neuroscience grew proportionately. It has now infiltrated into many areas, including philosophy, linguistics, artificial intelligence, and the other constituent cognitive sciences. One impression common by the end of the decade was that neuroscience, or at least a neuroscientific component, had entered in a substantive way into most research in the cognitive sciences—with two interesting exceptions.

It was clear that many mainstream branches of philosophy were getting along happily with no neuroscience component. One reason for this is that many philosophers adhere to the Wittgensteinian credo that the content of science in any form is irrelevant to philosophy. The two disciplines, science and philosophy, because of their different natures are distinct and discontinuous. Philosophy, it is held, is a purely conceptual enterprise, unaffected by the contingencies of empirical

1

research and data. Another philosophical position, which arrives at the same conclusion but is based in another tradition, holds that study of the human sciences (of which philosophy is one) is independent of the nature and methods of the natural sciences.

There is a second group that—while allowing science some relevance to the philosophical enterprise in general—nevertheless considers neuroscience to be too detailed, speculative, or both to provide the basis for claims that ought to be taken into account when philosophizing. Why, the argument goes, base philosophical theorizing on tentative results that will change within a few years? Philosophy may not be a discipline that is carried on *sub specie eternitatis,* but at least it should not be subject to the particular whims and fads that characterize a given science at a particular time.

More surprisingly, there is still considerable resistance on the part of the traditional psychological disciplines to integrating neuroscience into their agendas. As late as 1998, a review article in *American Scientist* concluded: "We did not find clear movement . . . toward integration of the new work in neuroscience with mainstream thought in psychology" (Robins et al. 1998, 310). This lack of influence in psychology (and, we might add, philosophy) contrasts with another finding the same authors made: "The core neuroscience journals are among the most frequently cited scientific journals when citations from all sources (not just psychology publications) are considered; in fact, two of the journals—the *Annual Review of Neuroscience* and *Trends in Neuroscience*—have citation rates comparable to *Science* and *Nature*" (312).

We do not feel competent to explain the lack of influence of neuroscience in psychology. In many cases with which we are familiar, psychologists who were once strict mental cognitivists have in fact modified their research programs to include a neuroscientific component. Often, however, the modification merely consists in the use of imaging techniques to study the problem of localization in the brain of a particular cognitive ability.

In philosophy the lack of attention to neuroscientific research is less surprising, given that many philosophers, as we noted previously, hold that science and scientific findings *tout court* are irrelevant to philosophy. But let us briefly address two ways in which such findings might be seen as relevant to philosophy.

First, and most controversially, neuroscience, and particularly cognitive neuroscience, tries to answer questions about perception, memory, and knowledge. These questions, in some form, touch the basic issues of epistemology.

Now some philosophers might hold that neuroscience, like any science, can only be descriptive and therefore is ill suited for normative or prescriptive philosophical inquiry. But this argument seems to miss two major points. First, the information gathered from neuroscience tells us about the cognitive systems that humans (and other animals) have and about how these systems function in knowledge acquisition and use. In many cases they challenge the basic assumption that epistemologists have made about the very nature of the "building blocks" of knowledge. For example, the object recognition system and the system for detecting spatial relations among objects seemingly are distinct from each other and also from the system that registers motion. From these findings it would seem that philosophers worried about our knowledge of the external world (to put the point in an old-fashioned way) might well attend to the different parameters of the world, and of the processing systems that carry and constitute knowledge for the knowing organism. So knowing how these systems function properly is a straightforward normative application of the science.

Furthermore, that there are different systems suggests that there are certainly more than two basic kinds of knowing (knowing how and knowing that). Many lines of research suggest strongly that for any kind of adequacy one must consider categorical (or conceptual) knowledge as different from spatial, and both of these as different from propositional (or, more generally, linguistically based) knowledge. Again, procedural knowledge comes to be established through experience in a different type of physical system than declarative or episodic knowledge. These systems have different properties (e.g., declarative systems are often said to have parts that may be isolated and recombined, whereas procedural systems cannot be broken down nor can their parts be used in different contexts). The systems function in radically different epistemic ways. And finally, we might mention that neuroscience has given us much new interesting information about how the body functions in representing and utilizing knowledge gained by perception of the world. This work on the kinesthetic and proprioceptive systems is crucial to understanding the forms of representation taken by our knowledge of the world. There is a strict phys-iological sense in which I am (or perhaps the self is) always in my perceptions and knowledge representations. To the extent that these claims are true, they provide normative constraints on any theory of knowledge.

We will leave untouched another major epistemological question, that dealing with the nature of consciousness and conscious experi-ence. Suffice it to say that neuroscience can claim many insights about the nature and function of consciousness. And some philosophers and neuroscientists have maintained that somewhere in the complex struc-ture of the brain lies the key to understanding the role of conscious experience. We think that in some sense this claim about consciousness probably must be true, but work on consciousness has thus far pro-duced only a morass of unsupported and ill-thought-out assertions. So we will avoid making any claims ourselves that might add to the confusion.

In any case these are not the ways of exploring the neurosciences that are specifically covered here. This volume deals with a second major way in which neuroscience is relevant to philosophy. The phi-losophy of neuroscience, like the philosophy of physics or biology, has much to teach us about the nature of science. The study of the nature of neuroscience is a branch of philosophy of science and so may be pursued in order that we may come to understand more broadly how science works. That many of the problems that cognitive neuroscience addresses are about knowledge, evidence used in reasoning, concept formation, and the like makes the study doubly interesting. This is what the current volume is all about.

The chapters in this volume are organized around a traditional set of themes in the philosophy of science. The book begins with a brief foray into the history of the discipline by Olaf Breidbach. Then the nature and structure of scientific theories, long a standard philosophical topic, is examined by Valerie Gray Hardcastle and C. Matthew Stewart with an eye to ferreting out what is distinctive about theories in the neuro-sciences. The nature of explanation and the question of how phenom-ena are to be explained have been a concern of philosophers since the time of Aristotle. Here Antti Revonsuo and Stephan Hartmann take on the problem of explanation and how explanations function in the neurosciences.

Another major problem in cognitive neuroscience is how cognitive abilities relate to neuroscientific mechanisms. Some philosophers and theorists have considered this problem to be one of reduction of the psychological to the physical, although it is seen by others as a question of finding causal relations and mechanisms among different types and levels of entities and events. William Bechtel surveys this problem and examines some specific cases and claims about this crucial relation. The topic is further studied, from the point of view of discovery, by Carl Craver and Lindley Darden. They examine how different parameters of the search for neuroscientific mechanisms drive research strategies, when one conceives of the general problem of discovery in this science as finding the mechanisms by which explanations of cognitive (and other) abilities are given.

The next set of chapters deals with specific aspects of the methodological problems of research and evidence in the neurosciences. Many techniques used in research have become essential and definitive of contemporary neuroscience. The use of the computer as a scientific research tool has opened up possibilities for simulation that have been crucial for theory building and that have supported the possibility and plausibility of specific models, theories, or mechanisms of how the brain works. Holk Cruse addresses the power and limitations of such simulation models. The computer itself has also provided us with a model of how neuroscientific mechanisms might work, suggesting the idea that the brain is a computational machine. But this raises the larger questions of what it means to be computational and what implications computational constraints have for theory building and modeling. Rick Grush develops a critique of one aspect of this problem: how the representational nature of the brain might be reconciled with demands for a computational model.

It is impossible today to imagine what neuroscience might be like without the new imaging techniques of PET and MRI (both nMRI and fMRI). A wealth of new data localizing functions and mechanisms have resulted from the widespread use of these noninvasive techniques. Yet they are not without their methodological problems, as James Bogen shows in his chapter. But such imaging is not the only research tool available. The use of animal models or model organisms has been most important in gathering data, both for comparative purposes and for directly understanding the mechanisms by which the human brain works. Kenneth Schaffner and Marcel Weber address some pressing questions that underlie this particular methodological ploy. Finally, clinical practice, observations, and tests still provide us with much fundamental evidence about how the brain functions. Paul-Walter Schoenle's chapter assesses ways in which clinicians can shed light on (and must make assumptions about) the nature of consciousness.

As should be clear from this overview, the authors in the volume address some of the most basic questions about theory, explanation, and evidence as they occur in the neurosciences. To our knowledge, no other book considers such questions in a systematic and general way. We therefore hope that both neuroscientists and philosophers will find insight and stimulation within these pages.

It remains only to acknowledge some important persons. We thank Martin Carrier, who, along with Peter Machamer, was responsible for putting the program together. Without the dedicated work of Gereon Wolters and his Konstanz collaborators in arranging things in Germany, the conference that led to the chapters included in this volume would not have been possible. In Pittsburgh, our thanks go to Karen Kovalchick, Joyce MacDonald, and Director Jim Lennox of the Center for the Philosophy of Science for attending to the final details of publication. The index was complied by Jacqueline Sullivan and Nicole Zeak, and we extend our gratitude to them.

REFERENCES

Crick, F. H. C. 1958. "On Protein Synthesis." In G. Pontecorvo et al., *The Biological Replication of Macromolecules*. Symposia of the Society for Experimental Biology 12. Cambridge: Cambridge University Press, 138–63.
Robins, R. W., S. D. Gosling, and K. H. Craik. 1998. "Psychological Science at the Crossroads." *American Scientist* 86: 310–13.

1

The Origin and Development of the Neurosciences

Olaf Breidbach

Institute for the History of Medicine, Science, and Technology,
Friedrich Schiller University, Jena, Germany

Neuroscience is not identical with brain research. The brain may be a fascinating object to study, even if one is not concerned with any functional attributions. However, neuroscience, precisely defined, deals with the functional organization and the resulting behavioral specifications of the brain. It describes the brain in terms of its basic functional principles, that is, the structure and reactions of those elements of which it is formed, namely the neurons (Shepherd 1988). Neuroscience deals with the functional morphology of brain tissue. Therefore, neuroscience cannot trace its experimental tradition back to Aristotle or to Renaissance researchers such as Pietro Pomponazzi.[1] Even Descartes did not offer a strictly neuroscientific concept (Carter 1983; French 1994; Des Chene 1996): he presented a conceptual framework to describe sensation and associations as mechanisms, putatively localized in specific subfractions of human tissue. The logical possibility of a mechanistic view, and not its details, was of interest in his account.

Neuroscience without Neurons?

In this chapter I argue that, on the one hand, concepts describing mental faculties are of the utmost importance for the origin and development of neuroscience, but that, on the other, these concepts are themselves not part of the history of neuroscience. Such relevant con-

7

cepts include the idea of a mathematical description of the structure of reasoning as proposed by Gottfried Leibniz, David Hartley's concept of an associative mind, Jean Condorcet's model of sensation, or the idea of an encyclopedic organization of knowledge put forward in the tradition of the *ars memorativa* (Breidbach 1996a, 1997). Obviously there is a difference between mind research and brain research. Brain research, however, has adopted concepts from mind research in order to understand what the brain is used for. Accordingly, to understand the conceptual basis of neuroscience, one must ask how far brain research is influenced by more ancient ideas about the mind.

Even at the outset of my chapter, I must admit some contradictions in my argument: On the one hand, I am restricting the history of neuroscience to the nineteenth and twentieth centuries, while, on the other, I refer to ideas that date back at least to the Renaissance. I do not mean to trivialize the historical background of brain research, which indeed traces its roots to antiquity. As has been shown by Clarke and Dewhurst (1972) and more recently by Finger (1994), it is only such a historical analysis that allows us to clarify the ideas that form the conceptual background of modern experimental research. However, as Brazier (1984) has outlined in her account of the origins of neurophysiology in the seventeenth and eighteenth centuries, there was indeed a shift in the methodological approach to brain research around 1800. Before that, general phenomena connected with the excitability of certain tissues were studied. Albrecht von Haller (1757–63), in the middle of the eighteenth century, was able to use such stimulations to arrive at a classification of various types of tissues. Nevertheless, it was not the functional organization of a tissue but its classification that was of interest (Neuburger 1897). Even in those studies that relied on a detailed morphological description of brain tissue, it was not a mechanism but certain characteristics of the excitability of these tissues that were described (Pourfour du Petit 1710).

That attitude changed at the start of the nineteenth century. Researchers, trained in analytical physiology and possessed of a detailed knowledge of brain anatomy, aimed for an analysis of the functional organization of the brain and not a mere morphological characterization of it. This, in my view, constituted the origin of neuroscience as we know it today: the situation changed, giving way to a new mood, one that sought to pose specific questions. Thus modern neuroscience, as Clarke and Jacyna (1987) have pointed out, was formed after 1800.

That is important because, in this new perspective, old facts were referred to without being disentangled from their former conceptual backgrounds. Accordingly, concepts are still active in neuroscience that were formed long before there was any idea about the functional organization of the brain. Such concepts predate any physiological characterization of mental faculties. Basic ideas about the organization of the mind were established before the emergence of neuroscience in the strict sense, that is, before an experimental approach to the mind became commonplace: the concept of an associative mind, of relevance to discussions of parallel distributed processing, originated at the time of Newton and Locke (Breidbach 1996a). Already at the end of the nineteenth century, neuronal networks were regarded as an adequate characterization of the functional organization of such an associative mind (Exner 1894; Breidbach 1999a).

Thus it can be shown that central concepts of modern neuroscience were in place at the end of the nineteenth century, that is, before modern physiological ideas about the neuronal organization of the brain had been developed. This situation will be described in greater detail for one of these concepts, that of neuronal networks, which as already mentioned was formed in the last decades of the nineteenth century (Breidbach 1997). There is a mismatch in timing between the development of concepts and the development of experimental techniques in the neurosciences, and it will be asked to what extent this disparity is of relevance to modern discussions of mind, brain, and cognition (Gardner 1985).

Stories about Soul and Brain

Neuroscience in the strict sense can be regarded as dating from around 1800. Medical descriptions of brain pathologies and even detailed dissections of the human brain—such as those published by Willis (1664)—were available in the seventeenth century. In the midst of the eighteenth century, medical researchers like François Pourfour du Petit (1710) had already described correlations between brain lesions and certain behavioral features; as has been mentioned, von Haller and his followers had put forward a physiological description that allowed them to define central attributes of neuronal tissue in physiological terms, and Julien Offray de Lamettrie (1748) had published his *l'Homme machine*. Nevertheless, the decades near the turn of the century brought something completely new. Not only that: starting with the work of Felice Fontana (1781) the histological structure of nervous systems became the object of microscopy studies; in discussions about animal galvanism, the physiological analysis of the nervous system was gradually placed within a completely new methodological framework (Shepherd 1991). In parallel, approaches that were interested only in defining the organ of the soul, and not in a more detailed study of its putative functional morphology, came to an end. The works of two men mark that change: Samuel Thomas von Soemmerring and Franz Joseph Gall (Hagner 1993).

In 1796 Soemmerring published a small treatise, *On the Organ of the Soul,* which proclaimed that the soul was to be found in the liquid of the brain ventricles. He explicitly described various features of brain morphology by studying the various brain nerves. For a functional interpretation, however, he referred neither to pathological studies nor to a detailed discussion of animal electricity; nor did he present experiments of his own on that subject, relying instead on a review of old authorities. What made this booklet especially interesting is that Soemmerring—somewhat naively—dedicated it to the philosopher Immanuel Kant, seeking a philosophical comment on what he thought to be the definitive anatomical answer in the quest to identify the seat of the soul.

Kant's comment, published as an appendix to Soemmerring's monograph, offered anything but philosophical support for the author's ideas. Referring to the kind of physiology pursued by Soemmerring, Kant proposed that Soemmerring should think a bit more about chemistry before developing such ideas. Even more importantly, Kant condemned any effort to define a specific organ of the soul, which—according to him—could not in principle be described physiologically. In consequence, physiology, according to Kant, should be reduced to a description of mechanisms of sensation, memory formation, and so on, revealing only the functionality underlying causal explanations of entities within the physical world. One had to refrain from imagining that such an analysis could go beyond the accessible, objective world, describable in purely physical terms.

Only a few years later, in his famous chat with Napoleon, Laplace declared that in his mechanistic interpretation of cosmogenesis he did not require hypotheses about the putative influence of God (Dieudonné 1985).

In the very decade in which Soemmerring published his account, Franz Joseph Gall in Vienna was permitted to give official lectures on his concept of the localization of mental faculties (Oehler-Kein 1990). Gall's concept, known as *phrenology*, was a mixture of highly qualified neuroanatomy and an ad hoc version of cognitive psychology. His revolutionary contribution to neuroanatomy was to identify certain brain areas not merely by external morphological criteria but by following the nerve pathways that connect the brain with the spinal cord and the sense organs. According to him the brain was not a functionally homogeneous but a diversified tissue (Breidbach 1997). Comparative anatomy showed, furthermore, that in the animal kingdom behavioral complexity was correlated with the expansion of certain brain regions. Since Gall's studies of hydrocephali had convinced him that during development the brain pressed on the skull, forming the morphology of its bones, he concluded that external skull features (bumps) were correlated with features of the underlying brain tissue. His comparative psychology regarded certain cognitive abilities as correlated with peculiarities in skull structure. As the latter was thought to be an epiphenomenon revealing the functional specification of the underlying brain tissue, Gall's idea was clear: he looked for specific cognitive abilities in humans correlated with peculiarities in skull structure. The outcome of this work was phrenology, based on the idea that measurement of skull anomalies in a given human being could reveal specific cognitive abilities.

This approach became very famous around 1800, as Gall, after being more or less sacked by the Austrian government and thus hindered in his academic career at home, organized a tour of Europe during which he privately lectured on his ideas (Mann 1985). Later he established himself in Paris, and—making use of the right to submit a previously unpublished monograph on a scientific subject for consideration by the Academie Royale—injected himself into contemporary academic discourse. Within a short time, however, his followers completely lost interest in his anatomical ideas, declaring—as George Combe did in 1833—that neuroanatomical knowledge would be useless for practicing phrenology.

Localization?

Following Gall's lectures and publications in the first decades of the nineteenth century, the idea of the neuroanatomical localization of cognitive functions was the subject of widespread discussion (Young 1970). In the 1820s, however, this discussion was brought to an end by the work of the French physiologist Flourens (1842). He described a series of lesion experiments, performed on animals of various groups, that seemed to establish that the cortex did not contain areas that are responsible for particular cognitive functions. Instead, according to Flourens, only the size of the brain might be responsible for the degree of behavioral complexity expressed by a certain individual. His writings—in which he quite uncritically compared various species and presented only minimal protocols for his experiments—deserve critical discussion on their own, but what is historically important is that his authority managed to discredit any theory of functional localization in the brain until the 1860s (Neuburger 1897). It was only by then that clinical neuropathology had been able to show that subjects became aphasic in cases in which a specific brain region in one hemisphere of the cortex was destroyed. At nearly the same time, physiologists showed that stimulation of certain brain regions resulted in defined motor movement patterns in dogs and apes (Fritsch and Hitzig 1870; Ferrier 1873; for review see Dodds 1878). Thus the argument of Flourens against a functional compartmentalization of the cortex was in time discredited by both neuropathology and electrophysiology.

One might reasonably guess that in the following decades the identification of ever smaller brain regions responsible for more specific behavioral patterns would have been advanced. In fact it was not until the middle of the twentieth century—when electrophysiological techniques had been developed that allowed the activities of single cells to be registered—that the identification of functional elements of brain tissue could be considered. Once these elementary characteristics of functional brain design had been established, more subtle questions— such as those pertaining to the superposition of different nervous pathways and the mechanisms of their association—could be dealt with.

Neuronal Networks

The attempt to define complex systems turned first to the study of much more simply structured model organisms (Strausfeld 1976), aiming for a nearly complete description of the functional elements of relevant tissues and their modes of combination. Such an approach seemed obvious—in spite of the fact that, in a strict sense, it is not adequate for biological research, because typological concepts should have been omitted with the rise of evolutionary biology. In a program in which different organisms are studied without reference to phylogenetic relations, neglect of an evolutionary perspective is implicit. Such an approach is marked by the notion of the model organism. In a model organism the basic principles of a particular physiological reaction can be studied and are interpreted to be of general relevance, irrespective of the phylogenetic position of the species. Such an approach is problematic, for it maintains a typological perspective. According to such a view, however, even the behavior of phylogenetically divergent systems (these are mere convergences in phylogenetic terminology) seemed comparable, so that, in its extreme, the British physiologist Edgar D. Adrian (1937) could argue that any analysis of the complexity of human electroencephalographic (EEG) patterns had to start by comparing the human EEG with the EEG of much simpler organisms such as beetles. By such a comparison of the simple (seen as elementary) with the more complex (seen as merely extending such elementary characteristics), it was thought to be possible to decode human brain activity patterns (Breidbach 1999b).

Up to the 1980s, the idea that complexity is in principle just a numerical extension of elementary characteristics provided arguments for comparing simple brains—which were smaller—with more diversified brains—which were bigger (Kuffler and Nicholls 1976; Huber and Markl 1983). Accordingly, the internal organization of the nervous system was defined in terms of such quantitative categories. Computation was thought to consist in sampling data in a hierarchically organized signal selection architecture (Creutzfeldt 1983). Through physiological techniques, so-called commander neurons were identified as top-level structures in these signal selection hierarchies (Kupfermann and Weiss 1978). These neurons were thought to be the devices by means of which a nervous system controls and maneuvers an organism.

Invertebrate organisms, with their much smaller brains, thus became favorite objects to study the principles of neuronal computation. In these organisms, neurons were individually characterized and identified. Because these cells allowed highly detailed analysis, their identification in the 1970s seemed to signal a definitive breakthrough in the analysis of the functional implications of wiring patterns in the nervous system. For some years there had been a search for multimodal properties in these cells, in the hope that they might allow an easy understanding of the way in which a multitude of functional circuits in the brain could work on a restricted number of neurons. In the 1980s, however, this optimism slowly dimmed; the identified neurons proved much more complex than had at first been thought.

Already in the 1970s, the analysis of neuronal plasticity had shown that neuronal couplings were not as fixed as had originally been thought when sensory pathways were described in purely physiological terms. Redundancy seemed to be a principal theme of neuronal coupling—especially in vertebrates. Analysis of the plasticity of sensory projection areas in the ape cortex and the projection patterns of vibrissae in rats had demonstrated the extent of functional plasticity in the wiring patterns of neuronal assemblies (Creutzfeldt 1983). At the same time, artificial neuronal networks provided new approaches for cognitive scientists working on artificial intelligence (Gardner 1985). Neuronal networks were thought to be suitable for simulating memory formation and pattern recognition and thus gave new impetus to discussions of cognitive architectures and the implementation of logical features in artificial intelligence (Palm 1982). By the end of the 1980s, the concept of neuronal networks had become even more attractive within the field of neurophysiology, because the fundamental design of such networks seemed to reflect basic neuroanatomical features (Hertz et al. 1994).

A more detailed view, however, suggests that the design of artificial neuronal networks bears only a slight resemblance to that of real neuronal networks, reflecting them only rudimentarily and even incorrectly. Not only is computation in such a network model reduced to a binary (or at best multivariate) reaction of the integrative units—a model that does not reflect the computational properties of a real neuron—but, even worse, the basic principle underlying real neuronal networks—the formation of a nonstatistically but locally expressed connectivity—is not reflected in artificial neuronal networks (Holthausen and Breidbach 1997).

So far my description has followed only one trail through the history of the neurosciences, ignoring biochemistry, developmental neuroscience, and clinical research. Thus this picture presents but a fragment of the complete story. Despite its straightforward way of looking at history, it is inaccurate. According to this view, neurobiology would have no need of a conceptual framework. Its commitment to dissecting the elementary reactions in neuronal functions, trying to identify the constituent elements of mental activities, would be sufficient to motivate research.

Throughout the twentieth century, neuroscience was quite effective in miniaturizing its perspective, scattering details and revealing more and more elements to study. Thus the dissection of functional pathways in the brain did not stop at the cellular level: it has progressed further—to the level of synapses and beyond to the level of individual membrane proteins. Until the late 1980s, there remained hope that, by increasing the degree of miniaturization, the basic design of a conscious brain could be disentangled, that its functional "atoms" could be dissected. Alexis Routtenberg (1989), famous for his molecular biological description of basic chemical mechanisms of the synapses, and Eric Kandel, who studied neuroethology on the basis of a selected set of snail neurons, argued to this effect at a meeting devoted to memory formation in 1988 (Kandel et al. 1989). This approach, however, did not ultimately yield the desired solution, even though it was effective in perpetuating the trend in neurobiological research of avoiding the mind and instead pursuing a continued miniaturization of perspective.

The kind of problems for the cognitive sciences that arose from such an approach can be seen in the reception within psychology of Kandel's results on sensitization of the specific behavioral feature he studied in the snail—the now-famous *Aplysia*. He found that a specific neuromodulator, serotonin, was important for input-controlled modification of synaptic transfer functions in the network that controlled the behavior he was interested in (Kandel 1976). On the basis of these findings he postulated the importance of serotonin for activity-dependent changes in the functional pattern of a neuronal network. He even discussed these mechanisms as a model for understanding some essential features in the formation of memory traces within the brain. Neuropsychologists, citing his work, did the reverse: they screened a population for anomalies in serotonin level, correlated such anomalies with memory deficiencies, and thereafter attempted to characterize memory formation defects by studying serotonin expression (Reichert 1990).

Neurons

What is the other part of the story? As I have explained, the functional anatomy of the brain was unclear in the first half of the nineteenth century (Loos 1967). The first fruitful attempt to dissect connectivities in the nervous system was directed to the nervous periphery: it was the detection of the reflex arc (Brazier 1988). This was achieved by studying the different properties of nerves, first in the head (Bell and Magendie) and later on in the frog leg system (Hall). In the end, sensory and motor nerves were discriminated and hypotheses about the functional wiring in the connecting neural tissue were proposed. Up to that point, however, apart from vague ideas about a putatively globular structure of the nervous system, no structural interpretation of the organization of the nervous system was available. In the 1850s Deiters described the structural complexity of a single nerve cell (Deiters 1865). However, the problem was that—prior to Rudolph Virchow's publication on cellular pathology (Virchow 1858)—understanding of the histology of vertebrate tissue was hindered by ignorance of the functional implications of the cellular structure of animal tissue (Dierig 1993). After 1860, due to improvements in histological techniques, neuroanatomy gained ground, even in regard to the microarchitecture of brain tissue. Important analyses were contributed by the Austrian neuroanatomist Theodor Meynert; Camillo Golgi invented a new staining method that provided new insights into the cellular architecture of nervous tissue. Supported by the developmental studies of Wilhelm His and the lesion studies of Auguste Forel, the neuron concept, which envisioned nervous tissue as composed of discrete nerve cells, seemed to have become definitively established by about 1880 (Breidbach 1993). Brain tissue was regarded as a wiring pattern consisting of discrete units with one-way connections, by means of which the function of the nervous system could be explained (Breidbach 1995a).

The neuroanatomist Meynert (1867–68) and the psychologist Alexander Bain (1868, 1875) proposed a functional organization of neuronal computation that made use of James Mill's concept of associative psychology (Breidbach 1996a). According to that concept, a certain sensation was transferred from the sensory organ via a sensory pathway into the brain. Because of the connectivity of neurons, the stimulations, depending on sensations of the corresponding neurons, would then disperse throughout the brain tissue, following the pathways laid out by the connections of the specific neurons. In effect, an oscillation is formed in the adjoining nervous tissue, promoted by the connected neurons (Mill 1869). The connections of neurons define their interactions. Sensory representations (the stimulations elicited in the nervous system via sense organs) are translated into a new code. Those stimuli that result in analogous firing patterns are internally regarded as identical; those that overlap to a certain degree in the elicited stimuli are regarded as analogous. Thus the superposition of neuronal activation patterns was interpreted as corresponding to those mechanisms described by Mill as making up the functional basis of putative associations: associations of the same type will elicit similar oscillations; a coupling of different sensation types will associate the respective sensory input situations, and so on. Thus the old concept of associative psychology, which Mill had adopted from David Hartley (1749), provided the framework for interpreting the functional morphology of nerve cell assemblies and, based on that, the functional organization of brain tissue.

According to Meynert, there were distinct brain regions in which such superpositions were computed. This hypothesis could be correlated with the data obtained for the functional compartmentalization of brain tissue by the earlier electrophysiological studies. Meynert's transformation of Mill's concept was adopted by those in the mainstream of late-nineteenth-century neuroscience (Breidbach 1995a). Indeed it can be regarded as the most widespread idea about the functional organization of brain tissue in the neuroscience of that time. Analogies between the functional organization of information flow in brain tissue and the functional organization of newly developed telegraphic equipment (hierarchy of information flow, wiring patterns) were obvious. As early as 1872, His (1893) launched a direct attack on scientists who tried to simplify the above-mentioned ideas about the functional organization of nervous tissue by uncritically adopting telegraphic techniques as a model for brain functions. As His explained, the presumed hierarchical organization of computational developments was missing in the real brain. He offered a new analogy—that of the Prussian bureaucracy—arguing that the monolithic picture of optimized function in that organization was a fiction, hiding less strict information-transfer functions. He argued that, to disentangle computational patterns in the brain, one must reconstruct the local connectivities of the different types of neurons.

The famous Spanish neuroanatomist Santiago Ramón y Cajal (1895) and the invertebrate neuroanatomist F. C. Kenyon (1896) established wiring diagrams of putative information flows in the nervous system based on histological findings. Such detailed information about putative local patterns of information transfer remained hypothetical, however, as long as the coupling characteristics between different neurons could not be unraveled directly by physiological techniques. Only by transferring the concepts proposed based on a physiological analysis of the peripheral reflex arc to the microdimensions of brain tissue could the idea of a unidirectional interneuronal connection be definitively established. This hypothesis seemed to be supported by the idea of the gating mechanism, provided by the synaptic coupling sites of neurons, that was suggested by histological data in the last decades of the nineteenth century (Hans Held's "Endknöpfchen"; Held 1897). Nevertheless, there were no physiological measurements to prove this assumption directly. The functional compartmentalization of the brain, put forward by physiologists on the basis of stimulation and deletion experiments, did not clarify the physiology of local wiring patterns, but instead established a hierarchy of physiological reaction pathways based on gross morphological scalings that was only loosely connected to actual histological data.

In the 1890s, the physiologist Sigmund Exner, who had acquired important data on the microcompartmentalization of the vertebrate brain by combining stimulation experiments and lesion studies, was able to attempt to describe the physiological basis of psychological phenomena, referring to anatomical data obtained by Meynert and his followers. In his monograph, published in 1894, Exner described the computational properties of the sensory system in terms of neuronal networks. He even offered a description of the organization of parallel distributed computation in a neuronal assembly and provided the first printed sketch of a neuronal network, one that is faithful to modern concepts even in its details. In his description of memory formation, he described how neuronal connections might be strengthened when a stimulated neuron is activated in parallel by another neuron. In such situations, Exner declared, the latter neuronal connection will be changed in such a way that further activation will pass thorough that specific neuronal connection more easily. Thus as early as 1894 Exner had introduced what was later to be called the Hebbian learning rule (Breidbach 1995b).

Why then, one must ask, did Donald Hebb (1949) himself have to reintroduce the rule a second time, half a century later? Exner had shown how far neurophysiology and neuroanatomy had proceeded to a common ground by 1900, allowing them to give a hypothetical but detailed account of local computational characteristics of the nervous system. Even more importantly, in his description Exner had explained anew the putative computational actions of the brain that Mill had proposed in his earlier account of associative psychology. Exner had described a neuronal mechanism that Mill was not able to introduce in the first decades of the nineteenth century. Exner had demonstrated the viability of Mill's approach to a mechanistic interpretation of the computational characteristics of the brain based on neurophysiology. One could argue that Exner solved the problem that Mill had faced at the beginning of the nineteenth century: Mill was able only to describe an abstract model for the functional mechanism of the human mind; Exner provided a physiological foundation. Accordingly, he could argue that he "explained" on the basis of a neurophysiological concept all the phenomena listed by Mill in his account. Exner's monograph not only described sensation and memory formation, it also described the neurophysiological mechanisms of falling in love, of free will, and even of aesthetic experience.

Exner is just one example illustrating this approach. Neuroscientists like Santiago Ramón y Cajal, Paul Flechsig, August Forel, or Carl Wernicke gave similar accounts (Breidbach 1997). Nevertheless, only ten years later that optimism seemed to have disappeared. What prevented these ideas from becoming part of the mainstream of ongoing neuroscientific research?

The facts are simple. The situation resulted from the failure of neuroanatomy to prove that the nervous system is an aggregation of distinct cells and not a syncytium. Accordingly, a functional morphological interpretation of neuroanatomical descriptions could not be given a proper basis (Breidbach 1993). When even Held, who had provided the first histological evidence of the existence of synapses, changed his position and agreed with those researchers who thought the neuronal tissue to be a syncytium, neuroanatomy was no longer able to lead the conceptual development of neuroscience (Breidbach 1996b).

Physiology was on its own. The black box of the brain could be dissected by increasingly refined experiments down to the level of single neurons. The problem, however, was that in such an experiment only one pathway for a particular stimulation on the propagating neuron could be described. The picture gained in combining different analyses was an additive one. Accordingly, concepts that tried to envisage the complexity of cellular interaction could not be grounded experimentally. Only in the late 1960s did the famous account of the cerebellum published by Eccles et al. (1967) demonstrate anew the efficiency of combinations of physiological and anatomical studies in disentangling the computational characteristics of the vertebrate brain. But even this magnificent account did not directly elicit a change in the basic attitude of neurophysiologists.

Around 1900, the starting point for the new, physiologically based approach to the brain was the concept of reflex action (Liddell 1960). Sherrington (1906) described the brain as a unit that integrates various reflex actions, which were thought to be the fundamental reactions of brain tissue.

So far, I have reduced this historical sketch to just a few aspects centered on the basic understanding of the functional mechanisms of neuronal computation. Neither clinical and behavioral studies nor the comparative anatomy of the cortex have been mentioned, because these seemed not to be important for the aspects of neuroscience described (for details see Finger 1994).

How central to the discussion of present-day neuroscience are the aspects of the development of the field just described? In my view they are quite central, because modern discussion of consciousness in cognitive science and in neurophilosophy is centered on the idea of parallel processes and the general question of a mechanistic interpretation of neuroscience (Gardner 1985; Churchland 1986).

In this chapter so far, it has been demonstrated that the concepts central to a modern discussion of neuroscience are themselves far from being modern; on the contrary, they date back to a period of natural history in which the neuron was not even known.

Architectures of Cognition?

Finally, I sketch some aspects of the history of the neurosciences that might reveal an even deeper intermingling of older philosophical concepts with modern neuroscience. The problem is that we must have in place concepts about the organization of the mind before we seek any physiological interpretation of its putative mechanisms. If it can be shown that even the forms in which we tend to represent the mind significantly antedate the origin of neuroscience, there may be serious consequences for a philosophical view of recent neuroscience.

One of the principal schemes that we tend to use to interpret the organization of the mind is the concept of localization. We tend to link mental processes to specific places in the brain. One may argue that pathology and stimulation experiments have in fact confirmed this. Yet a closer view will demonstrate that the truth is not that simple. In stimulation and inhibition experiments on the brains of patients (necessary as tests before brain surgery), the neurophysiologist G. A. Ojemann (1990, 1991) showed that specific features of language propagation are strictly localized; however, he also demonstrated that these loci are only temporary in a given individual and that they show great variation between different individuals.

My own modeling experiments demonstrating the physics of local computation in neuronal networks have shown that the idea of localization does not adequately address the specificities of structures that perform parallel computation (Breidbach 1996c; Holthausen and Breidbach 1999; Holthausen et al. 1999). The model tries to emphasize the idea that the "real" nervous system is defined by local connectivities, for example, the particular pathways defined by structurally distinct types of neurons. It can thereby be shown that a signal is not simply represented in the nervous system according to properties of the external world, but that it is computed based on internal system characteristics (Schmidt et al. 1996). Thus not merely the localization of a certain function in the network but instead its topological dynamics is computed. It can be shown that not a localization, but rather a topological registration employing spatial and temporal characteristics, is important for understanding the functional specifications of various nervous tissues (Holthausen and Breidbach 1997; Breidbach 1999c). To describe the latter, one must think in the mathematical terms of topology, that is, one must envisage a space-time continuum (Holthausen and Breidbach 1997). Such an approach does not contradict the results of localization studies—it extends them (Breidbach 1999b). The problem is not merely to demonstrate *where* but also to describe *how* something is being computed in brain tissues. Lesion studies, furthermore, show only how far—in the short run—normal behavior is disturbed after the brain is hurt in a specific way.

Modeling studies show that localization is not due to a fixed mapping, but that it is the result of a temporarily stable interrelation of different functional connections. This has been known for decades through experiments on the activity-dependent variation in size of the central sensory projection areas for tactile stimulation. Nevertheless, the idea of storing distinct functions in distinct places in the brain remains convincing. The question is whether this idea may have something to do with our concepts about the organization of cognition in general. In this case, the argument must go far beyond the putative physiological concepts, because the notion that things are stored in the mind in a precisely circumscribed locality is much older.

Architectures of the Mind

According to medieval thought, our insufficient logic may try to dissect the complexity of the world, but it will—because of its limited capacity—only scratch the surface of a deeper understanding (O'Meara 1982). Only a science that is able to reconstruct the ideas of God or at least the order in which He designed the terms used by logic might unravel something that is really true. Thus logic and, together with that, the mechanisms of the human mind are not sufficient for understanding what the world really is; their insufficiency is even an impediment to such understanding. Only when the real order of terms that reflects the order of the universe is presented can a knowledge be envisaged that is somehow sufficient. Accordingly, real knowledge would be an insight into the relationship of terms; knowledge thus would refer to "topic" (Schmidt-Biggemann 1983), meaning that it should describe the topology in which our ideas are placed.

With the *ars memorativa,* this speculative idea about the structure of knowledge had already found its technical realization in the medieval period (Berns and Neuber 1993). The *ars memorativa* dates back to the rhetoric of antiquity: in a culture that did not rely on written notes, memory had to be trained in such a way that it allowed the quick association of relevant ideas, quotations, and so on with a certain subject. The basic technique was to imagine an architecture. Relevant terms were imagined as placed in distinct locations within this architecture. Correlated ideas were envisaged as stored together in certain niches. The relative distances between particular terms in this imaginary architecture thus described their structural relations. Since they were associated according to their relative positions, the topology of terms in the architecture defined the structure of thought. According to this way of reasoning, definite truth must be found within an architecture of terms that is analogous to the structure that God originally formed for His ideas. In fact in the cabalistic traditions—both the Hebraic and the Christian—that idea had tremendous impact. For the Christian tradition, this influence can be traced to one author, Raimond Lullus, who in the thirteenth century had published an account of how to find that real order of terms (Yates 1964; Bonner 1996). His art cannot be outlined in detail here; in fact it is extremely complicated: Lullus presented something that we today might call an expert system. The point is that according to Lullus the localization in the architecture or internal room of the mind in which a term is fixed defines its meaning. Thus the real meaning of the name of something cannot be found in its being given a name, but rather in the relation that name has to other terms.

In the seventeenth century the Venetian senate paid a huge sum for a three-dimensional model of such an internal architecture based on Lullus's reasoning, in which each term had its distinct place (Schmidt-Biggemann 1998). It was said that by the relation each term had to every other term, its real meaning was directly intelligible. Accordingly, the topology of where the terms were found—their localization—yielded their real meaning.

Such a model gave access to a description of the internal organization of the brain. Conceived in such a way, intellectual training would aim at optimizing the construction of the internal architecture of the brain, so that the human microcosmos might reflect the external macrocosmos. The mind was to be regarded as an architecture. This concept is explicitly expressed by Robert Fludd (Yates 1966). According to his concept the mind is structured through its internal relations (Schmidt-Biggemann 1998); functionally it is seen as a storage room. The concepts of localization and fragmentation of mental faculties thus are the central themes in the *ars memorativa* and its cognitivistic interpretation. Their effectiveness can be studied in the educational program of the humanists, in which "topic" remained a central subject.

The concept of an architecture of the mind is thus part of an ancient tradition. In the seventeenth century, the first printed expert systems, encyclopedias (Alsted 1630), were structured according to the idea that any knowledge had to be referred to a topography of terms (Schmidt-Biggemann 1989). The system of terms, the structure of putative references of one term to another, was nothing other than a reflection of the conceptual framework of the *ars memorativa* (Florey 1993). The architecture of knowledge was designed to reflect a real architecture of the mind.

Space does not permit me to elaborate further on this tradition. What is important is that concepts of the hierarchy of mental faculties and the tendency to localize mental functions presented in the tradition of the *ars memorativa* significantly predate any discussion of localization patterns within the mind put forward in nineteenth- and twentieth-century neuroscience.

It is tempting to analyze to what extent the concepts described here have had an impact on our ideas about the constitution of the mind. That discussion is of even greater relevance, since with David Hartley the tradition of an associative concept of the mind dates back to the seventeenth century, that is, to a period in which the topological concepts sketched here were of relevance. Nevertheless, it remains to be seen to what extent the impact of such ideas was explicit in the nineteenth century. Terms like *natural system* were formed in that tradition (Knight 1981).

What is certain is that the idea of localizing mental functions was not first developed by neuroscience. The concepts of representation and localization were in place well before the framework of an experimental neuroscience was established (Schmidt et al. 1996; Ziemke and Breidbach 1996). The extent to which neuroscience really possesses a cabalistic touch, however, is a point for further discussion.

NOTES

I thank Paul Ziche for his critical comments on various versions of the manuscript.

1. For them, what was important was not the functional organization of the brain but the idea that a mechanism could be found by means of which human reactions could be explained. Thus Pomponazzi's (1990) commentary on the Aristotelian *de anima* was presented as an attempt to prove the mortality of the soul and thus to provide a naturalistic account of what a human being really is. Insofar as the core of his argument was a theological one, the putative functional organization of such a mortal soul was of interest only as such and not in its details (Gilson 1961; Di Napoli 1963; Pine 1973; Wonde 1994).

REFERENCES

Adrian, E. D. 1937. "Synchronized Reactions in the Optic Ganglion of *Dytiscus*." *Journal of Physiology* 91: 66–89.

Alsted, J.-H. 1630. *Encyclopaedia*. Herborn.

Bain, A. 1868. *The Senses and the Intellect*. 3rd ed. London: Longmans.

——. 1875. *The Emotions and the Will*. 3rd ed. London: Longmans.

Berns, J. J., and W. Neuber, eds. 1993. *Ars memorativa: Zur kulturgeschichtlichen Bedeutung der Gedächtniskunst 1400–1750*. Tübingen: Niemeyer.

Bonner, A. 1996. "Introduction." In *Raimundus Lullus. Opera*. Stuttgart: Frommann-Holzboog, 9–37.

Brazier, M. A. B. 1984. *A History of Neurophysiology in the 17th and 18th Centuries*. New York: Raven Press.

——. 1988. *A History of Neurophysiology in the 19th Century*. New York: Raven Press.

Breidbach, O. 1993. "Nervenzellen oder Nervennetze? Zur Entstehung des Neuronenkonzeptes." In E. Florey and O. Breidbach, eds., *Das Gehirn: Organ der Seele*. Berlin: Akademie-Verlag, 81–126.

——. 1995a. "Understanding the Functional Architecture of Cortical Tissue in the 19th Century." *Clio Medica* 33: 57–71.

——. 1995b. "From Associative Psychology to Neuronal Networks: The Historical Roots of Hebb's Concept of the Memory Trace." In N. Elsner and R. Menzel, eds., *Göttinger Neurobiologentagung 1995*. Stuttgart: Thieme, 58.

——. 1996a. "Vernetzungen und Verortungen: Bemerkungen zur Geschichte des Konzepts neuronaler Repräsentation." In Ziemke and Breidbach 1996, 35–62.

——. 1996b. "The Controversy on Stain Technologies: An Experimental Reexamination of the Dispute on the Cellular Nature of the Nervous System." *History and Philosophy of Science* 17: 3–30.

——. 1996c. "Konturen einer Neurosemantik." In S. J. Schmidt, G. Rusch, and O. Breidbach, eds., *Interne Repräsentationen: Neue Konzepte der Hirnforschung*. Frankfurt: Suhrkamp, 35–62.

——. 1997. *Die Materialisierung des Ichs: Eine Geschichte der Hirnforschung im 19. und 20. Jahrhundert*. Frankfurt: Suhrkamp.

——. 1999a. "Neuronale Netze, Bewußtseinstheorie und vergleichende Physiologie: Zu Sigmund Exners Konzept einer physiologischen Erklärung der psychologischen Erscheinungen." In S. Exner, ed., *Entwurf zu einer physiologischen Erklärung der psychischen Erscheinungen 1894*. Ostwalds Klassiker der exakten Wissenschaften, vol. 285. Frankfurt: Harri Deutsch, i–xxxviii.

——. 1999b. "Comparing Minds: A Comment on Beetles' Intelligence." *Theory in Biosciences* 118: 54–65.

——. 1999c. "Internal Representation: Prelude for Neurosemantics." *Journal of Mind and Behavior* 20: 403–20.

Cajal, S. Ramón y. 1895. *Les nouvelles idées sur la structure du système nerveux chez l'homme et chez les vertébrés*. Paris: Reinwald.

Carter, R. B. 1983. *Descartes' Medical Philosophy: The Organic Solution of the Mind-Body Problem*. Baltimore: Johns Hopkins University Press.

Churchland, P. 1986. *Neurophilosophy*. Cambridge, Mass.: MIT Press.

Clarke, E., and K. Dewhurst. 1972. *An Illustrated History of Brain Function*. Oxford: Sandfort.

Clarke, E., and L. S. Jacyna. 1987. *Nineteenth Century Origins of Neuroscientific Concepts*. Berkeley, Calif.: University of California Press.

Combe, G. 1833. *System der Phrenologie*. Braunschweig: Vieweg.

Creutzfeldt, O. D. 1983. *Cortex Cerebri*. Berlin: Springer-Verlag.

Deiters, O. F. K. 1865. *Untersuchungen über Gehirn und Rückenmark des Menschen und der Säugethiere*. Braunschweig: Vieweg.

Des Chene, D. 1996. *Physiologia: Natural Philosophy in Late Aristotelian and Cartesian Thought*. Ithaca, N.Y.: Cornell University Press.

Dierig, S. 1993. "Rudolf Virchow und das Nervensystem: Zur Begründung der zellulären Neurobiologie." In E. Florey and O. Breidbach, eds., *Das Gehirn: Organ der Seele*. Berlin: Akademie-Verlag, 55–80.

Dieudonné, J. 1985. *Geschichte der Mathematik 1700–1900*. Braunschweig: Vieweg.

Di Napoli, G. 1963. *L' immortalità dell' anima nel Rinascimento*. Torino: Società Editrice Internationale.

Dodds, W. J. 1878. "On the Localisation of the Functions of the Brain." *Journal of Anatomy and Physiology* 12: 340–660.

Eccles, J. C., M. Ito, and J. Szentágothai. 1967. *The Cerebellum as a Neuronal Machine*. Berlin: Springer-Verlag.

Exner, S. 1894. *Entwurf zu einer physiologischen Erklärung der psychischen Erscheinungen*. Leipzig: Deutiche.

Ferrier, D. 1873. "Experimental Research in Cerebral Physiology and Pathology." *West Riding Lunatic Asylum Medical Reports* 3: 30–96.

Finger, S. 1994. *Origins of Neuroscience: A History of Explorations into Brain Functions*. New York: Oxford University Press.

Florey, E. 1993. "Memoria: Geschichte der Konzepte über die Natur des Gedächtnisses." In E. Florey and O. Breidbach, eds., *Das Gehirn: Organ der Seele*. Berlin: Akademie-Verlag, 151–216.

Flourens, P. 1842. *Examen de Phrénologie*. Paris: Parlin.

Fontana, F. 1781. *Traité sur le venin de la vipère, sur les poissons américains, sur le laurier-cerise et quelques autres poissons végétaux. On y a joint des observations sur la structure primitive du corps animal. Différentes expériences sur la reproduction des nerfs et la description d'un nouveau canal de l'oeil*. 2 vols. Florence: Nyon.

French, R. 1994. *William Harvey's Natural Philosophy*. Cambridge: Cambridge University Press.

Fritsch, G. T., and E. Hitzig. 1870. "Über die elektrische Erregbarkeit des Grosshirns." *Archiv für Anatomie, Physiologie und wissentschaftliche Medizin* 37: 300–322.

Gardner, H. 1985. *The Mind's New Science*. New York: Basic Books.

Gilson, E. 1961. "Autour de Pomponazzi: Problématique de l'âme en Italie au début du XVIe siècle." *Archives d'Historie Doctrinale et Littéraire du Moyen Age* 28: 163–279.

Hagner, M. 1993. "Das Ende vom Seelenorgan." In E. Florey and O. Breidbach, eds., *Das Gehirn: Organ der Seele.* Berlin: Akademie-Verlag, 3–22.

Hartley, D. 1749. *Observations on Man, His Frame, His Duty and His Expectations.* London: S. Richardson.

Hebb, D. O. 1949. *The Organization of Behavior.* New York: Wiley.

Held, H. 1897. "Beiträge zur Struktur der Nervenzelle und ihrer Fortsätze." *Archiv für Anatomie und Entwicklungsgeschichte,* 204–94.

Hertz, J., A. Krogh, and R. G. Palmer. 1994. *Introduction to the Theory of Neuronal Computation.* Reading, Mass.: Addison-Wesley.

His, W. 1893. "Ueber den Aufbau unseres Nervensystems." *Verhandlungen der Gesellschaft Deutscher Naturforschung* 1: 39–67.

Holthausen, K., and O. Breidbach. 1997. "Self-Organized Feature Maps and Information Theory." *Network: Computation in Neural Systems* 8: 215–27.

——. 1999. "Analytical Description of the Evolution of Neural Networks: Learning Rules and Complexity." *Biological Cybernetics* 81: 165–75.

Holthausen, K., M. Khaikine, and O. Breidbach. 1999. "Neural Computation with Ideal Approximation Elements." *Theory in Biosciences* 118: 113–24.

Huber, F., and H. Markl, eds. 1983. *Neuroethology and Behavioural Physiology.* Berlin: Springer-Verlag.

Kandel, E. R. 1976. *Cellular Basis of Behavior: An Introduction to Behavioral Neurobiology.* San Francisco: W. H. Freeman.

Kandel, E. R., D. Glanzman, N. Dale, T. E. Barzilai, T. E. Kennedy, and S. Sweatt. 1989. "A Molecular Biological Approach to Long-Term Memory in *Aplysia.*" In H. Rahmann, ed., *Fundamentals of Memory Formation: Neuronal Plasticity and Brain Function.* Stuttgart: Fischer, 310–23.

Kenyon, F. C. 1986. "The Brain of the Bee." *Journal of Comparative Neurology* 6: 133–210.

Knight, D. 1981. *Ordering the World: A History of Classifying Man.* London: Burnett.

Kuffler, S. W., and J. G. Nicholls. 1976. *From Neuron to Brain.* Sunderland, Mass.: Sinauer.

Kupfermann, I., and K. R. Weiss. 1978. "The Commander Neuron Concept." *Behavioral and Brain Science* 1: 3–39.

Lamettrie, J. O. de. 1748. *l'Homme machine.* Leyden: Luzac.

Liddell, E. D. T. 1960. *The Discovery of Reflexes.* Oxford: Clarendon Press.

Loos, H. van der. 1967. "The History of the Neuron." In H. D. Hydén, ed., *The Neuron.* Amsterdam: Elsevier, 1–48.

Mann, G. 1985. "Frans Joseph Galls kranioskopische Reise durch Europa 1805–1807." *Nachrichtenblatt der Deutschen Gesellschaft für Geschichte der Medizin, Naturwissenschaften und Technik* 34: 86–114.

Meynert, T. 1867–68. "Der Bau der Grosshirnrinde und seiner örtlichen Verschiedenheiten, nebst einem pathologisch-anatomischen Corollarium." *Vierteljahresschrift für Psychiatrie* 1: 77–93, 125–217, 381–402; 2: 88–113.

Mill, J. 1869. *Analysis of the Phenomena of the Human Mind.* London: Longmans, Green, Reader and Dyer.

Neuburger, M. 1897. *Die historische Entwicklung der experimentellen Gehirn- und Rückenmarksphysiologie vor Flourens.* Stuttgart: Enke.

Oehler-Klein, S. 1990. *Die Schädellehre Franz Joseph Galls in Literatur und Kritik des 19. Jahrhunderts.* Stuttgart: Fischer.

Ojemann, G. A. 1990. "Organization of Language Cortex Derived from Investigations during Neurosurgery." *Seminars in Neuroscience* 2: 297–306.

———. 1991. "Cortical Organization of Language." *Journal of Neuroscience* 11: 2281–87.

O'Meara, J., ed. 1982. *Neoplatonism and Christian Thought.* Albany: State University of New York Press.

Palm, G. 1982. *Neural Assemblies: An Alternative Approach to Artificial Intelligence.* Berlin: Springer-Verlag.

Pine, M. L. 1973. "Pietro Pomponazzi and the Scholastic Doctrine of Free Will." *Revista Critica di Storia della Filosofia* 28: 3–27.

Pomponazzi, P. 1990. *Abhandlung über die Unsterblichkeit der Seele,* transl. and ed. B. Mojsisch. Hamburg: Meiner.

Pourfour du Petit, F. 1710. *Trois lettres d'un médecin des hôpitaux du roy.* Namur.

Reichert, H. 1990. *Neurobiologie.* Stuttgart: Thieme.

Routtenberg, A. 1989. "Role of Protein Kinase C and Protein F1 GAP 43 in Presynaptic Terminal Growth Leading to Information Storage." In H. Rahmann, ed., *Fundamentals of Memory Formation: Neuronal Plasticity and Brain Function.* Stuttgart: Fischer, 283–98.

Schmidt, S. J., G. Rusch, and O. Breidbach, eds. 1996. *Interne Repräsentationen: Neue Konzepte der Hirnforschung.* Frankfurt: Suhrkamp.

Schmidt-Biggemann, W. 1983. *Topica Universalis.* Hamburg: Meiner.

———. 1989. Vorwort. In J. H. Alsted, ed., *Encyclopaedia,* vol. 1. Stuttgart: Frommann-Holzboog, v–xviii.

———. 1998. *Philosophia perennis: Historische Umrisse abendländischer Spiritualität in Antike, Mittelalter und Früher Neuzeit.* Frankfurt: Suhrkamp.

Shepherd, G. M. 1988. *Neurobiology.* New York: Oxford University Press.

———. 1991. *Foundations of the Neuron Doctrine.* New York: Oxford University Press.

Sherrington, C. S. 1906. *The Integrative Action of the Nervous System.* New York: Scribner.

Soemmerring, S. T. 1796. *Über das Organ der Seele.* Königsberg: Nicolovius.

Strausfeld, N. 1976. *Atlas of an Insect Brain.* Berlin: Springer-Verlag.

Virchow, R. 1858. *Die Cellularpathologie in ihrer Begründung auf physiologische und pathologische Gewebelehre.* Berlin: Hirschwald.

Von Haller, A. 1757–63. *Elementa Physiologiae Corporis Humani.* 5 vols. Lausanne: Bousquet.

Willis, T. 1664. *Cerebri Anatome: Cui Accessit Nervorum Descriptio et Usus.* London: Martyn.

Wonde, J. 1994. *Subjekt und Unsterblichkeit bei Pietro Pomponazzi.* Leipzig: Teubner.

Yates, F.A. 1964. *Giordano Bruno and the Hermetic Tradition.* London: Routledge.

———. 1966. *The Art of Memory.* London: University of Chicago Press.

———, ed. 1982. *Lull and Bruno: Collected Essays,* vol. I. London: Routledge.

Young, R. M. 1970. *Mind, Brain and Adaptation in the Nineteenth Century: Cerebral Localisation and Its Biological Context from Gall to Ferrier.* Oxford: Clarendon Press.

Ziemke, A., and O. Breidbach, eds. 1996. *Repräsentationismus: Was sonst?* Braunschweig: Vieweg.

2

Theory Structure in the Neurosciences

Valerie Gray Hardcastle
Departments of Philosophy, University of Cincinnati, Cincinnati, Ohio, and
Virginia Polytechnic Institute and State University, Blacksburg, Virginia

C. Matthew Stewart
Department of Otolaryngology, University of Texas Medical Branch, Galveston, Texas

The title of this chapter is a bit of a misnomer. At least, we are not
going to be talking about theories in a traditional sense. Philosophers
generally imagine theories to be neat, tidy, simple things, preferably
able to be expressed with equations, like $E = mc^2$. We do find equations
in neuroscience—the Nernst equation describing the influx and efflux
of potassium and calcium across the cell membrane is one—but neat,
tidy, and simple do not describe neuroscientists' hypotheses. We are
not sure that most philosophers would recognize what we are going to
be calling a theory in neuroscience as a scientific theory at all—perhaps
"theoretical framework plus lots of details" would be a more accurate
description. Be that as it may, this chapter will briefly describe how
neuroscientists structure their understanding of the central nervous
system (CNS) in order to investigate it profitably.

We already know that brains are complicated and messy affairs; we
are here to tell you that theories about brains share these same traits.
The difficulty is that, in order to form a simple generalization about
how, say, the brain can reshape itself following insult or injury, scientists must retreat to such broad level of abstraction that their assertions
become almost meaningless empirically. In order to make their claims
testable in a laboratory, neuroscientists must confine their ideas to
particular animals, to particular experimental tasks, or to both. As a
result, we end up with neuroscientific "theories" that contain two
distinct parts: a broad statement of theoretical principle and a set of

detailed descriptions of how that principle plays out across different animal models and experimental tasks. Though the detailed descriptions fall under the general principle, they are not immediately derivable from it. Moreover, the detailed descriptions can be incompatible with one another, though each will maintain a family resemblance with the others. (For similar approaches to understanding theories in the biological sciences, see Schaffner 1980, 1993; Suppe 1989; Hardcastle 1995.)

We will be discussing recent research in neural plasticity to bolster our claims, though we believe that all areas of neuroscience fit under our analysis. As in much of neuroscience, work discerning the principles behind plasticity relies on several different animal models, from the frog to the gerbil to the cat to primates and humans. And it uses experimental protocols geared toward the specific properties and traits of the different animals. We can train monkeys to do more sophisticated things than frogs, but frogs are much cheaper and easier to procure, maintain, and sacrifice. Also, as in much of neuroscience, scientists interested in plasticity and learning aim to fill in the neural black box that psychologists had left hanging between stimulus input and behavioral output. Plasticity is a particularly odd phenomenon for neuroscience because it occurs so rapidly—we can recover function much faster than we can grow new synaptic connections, for example—but we have few clues about how the brain recovers or compensates for lost function.

Animal Models: A Study in Differences

At a gross level, mammalian brains are remarkably similar to one another. Indeed, the CNS in invertebrates is not all that different from the mammalian CNS. We find innumerable homologous areas, cell types, neurotransmitters, peptides, chemical interactions, and so forth. However, once we scratch the surface of different animal brains, we do find important differences.

Semicircular Ear Canals and Other Oddities

On the one hand, all mammals have roughly the same five end organs in their ears to support their auditory and vestibular systems. On the other hand, mammals work to keep the lateral semicircular canals in their ears parallel to the horizontal plane relative to the earth, for keeping them in that position allows mammals to obtain the best possible information about head position in space. The lateral canal is maximally excitatory to a yaw (left to right) head motion; keeping the canal in line with the horizontal plane allows the organ to detect this motion with the greatest accuracy. Rodents ambulate with their necks extended, which keeps their heads in an extreme dorsal position. Humans, in contrast, incline their heads about twenty degrees when walking naturally. Despite the extreme differences in how these mammals hold their heads, in both cases the lateral semicircular canal in the ear remains parallel to the horizontal. In general, we can correlate the differences in the shape of the semicircular canals in the ear with skull shape and the normal position of an animal's head. (It is an unanswered but intriguing question whether we find the canal structures we do because different heads evolved to be oriented in different directions or whether animals naturally hold their heads in different positions because their semicircular canals evolved differently.)

There are additional striking differences between herbivores and predators in brain structure, for creatures who munch on grasses and trees require much less precise environmental information than those who hunt moving targets in order to survive. Rodents, for example, have no foveae. To maintain visual fixation on a point, they move their necks, using what is known as the vestibular-colic response. The vestibular system in their ears tells them how their heads are oriented, and they use that information to reorient their heads in order to keep whatever object currently fascinates them in their line of sight.

In contrast, primates have foveae, and we move our eyeballs to keep our target within the foveal area, using the vestibular-ocular response. This is a much more precise orienting mechanism that allows us to move our eyes to compensate for changes in head position such that we can keep objects foveated for as long as we wish. For some indication of how important computing horizontal eye motion is to our brains, consider that the abducens (or VIth cranial) nerve in humans, which controls horizontal eye abduction, feeds into one of the biggest motor nuclei in the brain stem. This ocular nucleus, which controls only one very tiny muscle, is only slightly smaller than the nucleus that controls all of our twenty or so facial muscles.

In more striking contrast still, bats do not maintain ocular position in the same fashion as the rest of the mammals. Because they fly and so have greater freedom to move in three-dimensional space, maintaining body position relative to the horizontal is not an easy option. As a result, they use other sense organs, primarily hearing (the other half of the VIIIth cranial nerve), to determine how their eyes should be oriented. Consequently, they need not rely on vestibular-ocular responses as we do, even though their bodies are equipped with such reflex machinery.

All of these anatomical and physiological differences are important when neuroscientists want to investigate a subject like how the brain learns to compensate for damage to the vestibular pathways. What may seem small and insignificant differences from a broad mammalian perspective become hugely important as scientists seek to understand the particular mechanisms of brain plasticity. Can they use animals with no foveae and a vestibular-colic response to learn about how foveated mammals recover their vestibular-ocular response? More generally, how well do particular animal models translate across the animal kingdom? Should we be allowed to generalize from experiments on a single species (or set of species) to how nature functions? Let us look at a specific example to get a sense of how we should answer these questions.

Static Deficits with Canal Lesions

In all vertebrates, a unilateral labyrinthectomy (UL), or a lesion of the labyrinthine structure in one ear, gives rise to two types of ocular motor disorders: static deficits, such as a bias in favor of looking toward the lesioned side when the head is not moving, and dynamic deficits, such as abnormal vestibular-ocular reflexes (VOR), which occur in response to head movements. Within only two or three days following the UL procedure, the brain starts to compensate for its loss and the static deficits disappear. Since labyrinthine structures do not regenerate, and peripheral neurons continue to fire abnormally, whatever the brain is doing to recover has to be a central effect (Schaefer and Meyer 1974). Single-neuron recordings from a variety of animals indicate that the vestibular nuclei (VN) on the same side of the brain as the lesion start to show normal resting rate activity as the brain learns to compensate for its injury. The exact mechanism the brain uses for vestibular compensation is still unknown. Studies have shown that the brain does not substitute other sensory inputs or increase inputs from central regions that process sensory inputs, nor does it generate new synapses. A few data support the hypotheses that denervation supersensitivity or rapid hormonal changes could lead to vestibular compensation, but the effects of both appear to be small (see Smith and Curthoys 1989 for review).

However, scientists do believe that, whatever the mechanism is, it is also likely to be a general procedure the brain uses for recovery, for we find similar resting rate recoveries of the sort we see with the ipsilateral VN following denervation in the lateral cuneate nucleus, the trigeminal nucleus, and the dorsal horn, among other areas (Loeser et al. 1968; Kjerulf et al. 1973). Neuroscientists argue that the phenomenon of vestibular compensation should serve as a model for studying brain plasticity in general. That is, they argue that they are justified in abstracting from experiments on a single species or a single system to explain how brains work in general.

For example, Miles and Lisberger (1981, 296) conclude that "the VOR is a particularly well-defined example of a plastic system and promises to be a most useful model for studying the cellular mechanisms underlying memory and learning in the central nervous system." In their review of vestibular compensation, Smith and Curthoys (1989, 174) concur that "the recovery of resting activity in the ipsi VN following UL is an expression of a general CNS process which functions to offset the long-term changes in tonic synaptic input which would otherwise be caused by denervation" (see also Galiana et al. 1984). The scientists believe, at any rate, that the physiological and other differences we find across animals do not affect the theoretical claims they want to make about neural plasticity in the brain.

Exactly how an argument to defend their convictions is supposed to run, though, is unclear, since it is fairly easy to find significant differences in how organisms recover and compensate for vestibular damage across the animal kingdom. Frogs, for example, appear to rely on input from the intact labyrinth to regulate the resting activity of the VN. Mammals, however, do not. The recovery of their VN occurs independent of transcommissural inputs (Flohr et al. 1981). In addition, static symptoms follow different time courses in different animals. In rats spontaneous nystagmus disappears within hours after UL, whereas in the rabbit and guinea pig it persists for several weeks (Schaefer and Meyer 1973; Baarsma and Collewjin 1975; Sirkin et al. 1984). In humans, it may continue in one form or another for several years (Fisch 1973).

Needless to say, neuroscientists are not blind to these facts. In the same article, Miles and Lisberger (1981, 274) caution that "species differences concerning the detailed mechanisms are beginning to emerge." Their solution is to "avoid confusion" by looking primarily at work on the monkey (274–75). In their review article, Smith and Curthoys (1989, 162) themselves note that "the differences in VN neuronal activity between the cat, guinea pig, and rat suggest that some aspects of the neuronal changes underlying vestibular compensation may be species-specific even within mammals. This result would not be too surprising given that the speed of behavioral recovery differs fairly substantially between species."

Tensions in Neuroscientific Theorizing

One may well begin to wonder whether these scientists should succumb to cognitive dissonance, claiming both that they seek a general understanding of brain plasticity based on studies of one system from a few species and that differences across animals affect how brains are plastic in their exemplar cases. In short, it is difficult to square the notion that vestibular compensation is an instance of a general CNS phenomenon with the claim that vestibular compensation is species specific, especially when these two statements occur together in the same articles.

It is difficult, we agree, but it is not impossible. The scientists are not being completely irrational here. They hope that if they abstract across species differences, a common mechanism for neuronal plasticity will remain. Of course, if they focus on the individual species, they can point to specific differences in how that mechanism functions across the animal kingdom. Nevertheless, they are looking for a commonality across differences.

But this solution is not quite as neat as it sounds, for there is a fundamental tension in neuroscience between the big-picture story and what we find in particular instances. All sciences strip away features of the real world when they devise their generalizations. Physicists neglect friction; economists neglect altruism; chemists neglect impurities; and so on. However, what neuroscientists are doing is not analogous with what the physicists, economists, and chemists are doing. In each of the other cases, the scientists are simplifying the number of parameters they must consider in order to make useful and usable generalizations.

In contrast, if neuroscientists were to ignore the differences they find across species, they would have no data left with which to build a theory. There is nothing left over, as it were, once neuroscientists neglect the anatomical and physiological differences found in the brain across the animal kingdom. There is much left over when physicists neglect friction; most of classical mechanics is left, in fact. In distinction to the other sciences, we find a tension in neuroscience between the general rules one hopes to find that describe all brains and the particular cases neuroscientists happen to study. This is a tension that cannot be easily resolved.

Theory Structure in Neuroscience

What should the scope and degree of generalization for neuroscientific theories be? It appears we are confronted with an unpleasant choice. Either we settle for large-scale abstract generalizations, which gloss over what may be important differences, or we focus on the differences themselves, at the expense of what may be useful generalizations. Neither option is ideal; neuroscience needs a third path.

We believe that neuroscience has one available. Despite appearances, we do not have an either-or proposition before us that we must resolve before we can move ahead. For a proper neuroscientific theory contains both general (and fairly vague) abstractions as well as detailed comments on specific anatomies and physiologies. The paradigm theories for physics are simple, elegant equations with universal scope. Theories in neuroscience read more like a list of general principles plus detailed commentaries. This is the same difference that we find between Gödel's incompleteness theorem and the Torah with all its associated rabbinical remarks. Both contain wisdom, but they contain it in different ways.

One feels the tug of the dilemma only if one is operating with a restricted notion of what a scientific theory is. Some theories are pithy and succinct; some are not. Neuroscientific theories are not. A specific example should help make our claims clearer.

Goldberger (1980) divides the sorts of compensation a brain can perform following a lesion into three categories. There is *sensory substitution,* in which different sensory receptors trigger the same behavior as occurred before the operation or injury; *functional substitution,* in which the precise neural pathways differ, but the same neural subsystem is used as before; and *behavioral substitution,* in which the organism devises an entirely new behavior that was not originally part of its normal repertoire. These are very general observations concerning which strategies the brain might use in reorganizing itself. It is a theoretical description at the most general level; it is what we might call the "theoretical framework"—the most general component in a neuroscientific theory.

Once we adopt the framework, we can formulate more precise hypotheses as a way of filling out our theoretical proposal. Given the preceding discussion, we can rule out behavioral substitution as a description of vestibular compensation. The data suggest that recovery from the static symptoms occurring with UL probably results from functional substitution. The frog uses the same system on the other side of its head to force the VN back to normal resting activity levels. These observations can be local to particular phyla or species; hence they are not intended to serve as a more detailed specification of the general framework. Instead, they can be thought of as instances or examples of how the framework might be cashed out in particular cases.

However, it is not the case that all "fillings out" fail to generalize. For example, the dynamic symptoms of UL recover using a different mechanism (probably). One hypothesis is that brains use a form of sensory substitution to compensate for the vestibular-ocular reflex (Miles and Lisberger 1981; Berthoz 1988). In this case, the brain uses internally generated signals from the visual or somatosensory systems to compensate for the vestibular loss. It may substitute computations from the saccadic or a visual pursuit system, both of which (probably) reconstruct head velocity internally, for vestibular throughputs. Data drawn from experiments on frogs, cats, and humans indicate that they all apparently use the same mechanism, though it remains to be seen whether this proposal will be applicable to all creatures and whether it can be generalized much beyond vestibular reflexes.

Our intention here is not to advocate this hypothesis so much as to point out that there are different degrees of abstraction one might use once some theoretical framework has been adopted. Some discussions are going to be restricted to a single species, or maybe even one developmental stage within a species; others will include several unrelated species or phyla. Both are legitimate ways of cashing out the framework in particular instances, and neither is to be preferred over the other. The data will dictate the scope of subhypotheses, and scope can vary dramatically.

Experimental Protocols: A Source of Incomplete Information

The picture of neuroscientific theories we are developing is further complicated by the fact that legal, ethical, technological, and financial considerations restrict the sorts of experiments neuroscientists can run. As a result, they can conjoin the broad generalizations they make with only specific animal models. Hence, theories in neuroscience become a dialectic between general frameworks and skewed data sets, with each acting to inform the other.

The Dialect in Gathering Data

Scientists commonly use two types of impairments to analyze cognitive brain function. They can either lesion some part of the CNS, so that the animal must now work with a dysfunctional brain, or alter stimuli inputs, so that the animal must somehow compensate for degenerate data. Sometimes they do both in the same experiment. Neuroscientific experiments range from the extremely invasive to simply observing normal behavioral responses, and no one protocol is suitable for all animals. Not surprisingly, it is more difficult to perform invasive procedures on some animals (humans) than others (sea slugs), just as it is more difficult to fiddle with normal motor responses in some creatures (lobsters) than others (primates). Scientists have to tailor their investigations to the animals they are using. Not only must neuroscientists formulate theories that cover different anatomical structures having different physiological responses, but they also must cope with fundamentally incomplete data sets.

To investigate vestibular compensation under dynamic conditions in particular, neuroscientists must study how animals behave as their heads rotate back and forth. There are commonly two paths to take in this investigation. One is to perform single-cell recordings on anesthetized UL animals whose heads have been tied to a machine that moves back and forth in a sinusoidal motion (a little like being tied to a record player as a DJ scratches the record). The difficulty with this approach is that anesthesia can damp down neuronal responses so that the brain is no longer responding as it would were it awake. The second approach is to decerebrate animals and then tie their heads to the motion machine while they are awake. Removing large parts of animals' brains can also contaminate results, for, obviously, the animals are missing many, many inputs to the brain stem areas. However, it is necessary to remove the cerebellum so that awake animals can remain still enough for clean single-cell recordings to be obtained and so that motor planning inputs will not skew the neuronal firing patterns. Ideally both experiments are done and the results are used to triangulate in on how the brain behaves under more normal conditions.

Even more ideally, these experiments would be performed on primates so that we could discover how *our* vestibular system works. But humans and monkeys are much too expensive and difficult to keep for scientists to use them in this manner, ethical considerations aside. Neuroscientists must be able to run dozens of these experiments to guarantee accuracy in their data. Consequently, the inexpensive and pervasive gerbil, guinea pig, and cat are favorite animal models. Though neuroscientists do uncover features of recovery particular to those populations, they also use these models to support more general conclusions, which serve as a guiding framework for less invasive studies in other mammals.

For example, Newlands and Perachio (1990a,b) draw two conclusions from their single-cell recordings in UL decerebrate gerbils: one specific to their preparation and one a more general theoretical principle. First, they conclude that vestibular compensation for dynamic responses in the gerbil is due to inhibitory commissural connections running from the intact to the lesioned side, even though the connection strength between left (lesioned side) type I neurons and right (intact side) type II neurons substantially decreases. But they explicitly note that these data do not dovetail with similar experiments performed on the guinea pig or cat (1990b, 377–78). Perhaps other mammals do in fact work similarly; differences in experimental design could explain the discrepancies between their results and those of others. Nevertheless, the differences in design or physiology do not prevent them from drawing their second and more general conclusion: that recovery occurs using only the original populations of type I and type II neurons; no other group of cells steps forward to assume the responsibilities of what these neurons are supposed to be doing.

Although their specific conclusion concerning gerbil physiology is hard to generalize to other studies in neural plasticity, their more general conclusion is useful in guiding research using different models. If the same neurons in the VN are used in recovery, then they must be getting orientation information from somewhere other than the (now ablated) semicircular canals. If recovery occurs too quickly for the growth of new connections to explain how our brains are compensating for their loss, then some other sensory system must already be feeding into the vestibular system. One possibility, mentioned previously, is our visual system. Perhaps, as animals try to orient toward targets, error signals from the retina help the vestibular system compute head location. If this is the case, then disrupting visual processing in the brain stem should also disrupt vestibular compensation. Work in Perachio's laboratory is currently directed toward answering this question. Perachio and his colleagues have lesioned the first relay in horizontal field motion processing in the brain stem in monkeys trained to fixate on visual targets under a variety of conditions.

Perachio et al. are now asking whether this sort of lesion makes any difference to vestibular compensation, and, if so, what sort of difference it makes. Notice that they must switch to primates to answer this question, even though gerbil research originally inspired the inquiry. Remember: gerbils are rodents, so they have no foveae. Even if they had the intelligence to be trained to fixate on command, they do not have the physiology to do so. A predator is required for this task.

Presumably this research will uncover physiological connections specific to foveated mammals (or perhaps only to primates). If the research project turns out as expected, then we will learn that the ocular nucleus in the brain stem influences the vestibular system in macaque monkeys. However, the research would still fall under Newlands and Perachio's original guiding principle, that no new cell systems are involved in vestibular compensation. When the semicircular canals become damaged or input to them becomes attenuated, the brain increasingly relies on information from visual error detection to compute head and eyeball location. At the same time, the research will also advance a new guiding principle for neuroscience: that sensory systems in the brain overlap. If Perachio and his colleagues are right, then our visual and vestibular systems are fundamentally intertwined. Our eyes help tell our ears how our head is moving. Perhaps other sensory systems in the brain are similarly interconnected.

The Dialectic in Theorizing

And this is how theories in neuroscience are built and structured. Detailed conclusions regarding a single animal model give rise to general theoretical principles. These principles inspire new experiments with other animal models, which in turn give us new (and probably incompatible) details but also new general principles. These new principles then connect to other detailed studies using different protocols on still other animals, and so it goes.

We can take Perachio's new guiding principle that sensory systems overlap and connect it with other studies that investigate how brains compensate for injury, but that use different animal models and different experimental protocols. To wit: suspecting that our brain systems overlap dovetails with studies of monkey cortex. When the median nerve of the hand is severed in adult owl or squirrel monkeys, areas 3b and 1 in somatosensory cortex (areas that normally respond to medial nerve stimulation) begin responding to inputs from other nerves in the hand (Merzenich et al. 1983; see also Clark et al. 1988; Barinaga 1995). Merzenich argues that, given the rapidity of the response, silencing the radial nerve inputs "unmasks" secondary inputs from other afferent nerves. At first, areas 3b and 1 only crudely represented the dorsal surfaces of the hand. Over time, they transformed into highly topographic representations. A month after surgery, the areas that formerly responded to the median nerve now completely responded to the alternative inputs in standard hypercolumn form. Apparently, using mechanisms still unknown, the brain capitalizes on previous secondary connections to compensate for lost inputs. (Here again, though, once we move beneath the general story, interesting differences emerge. Even within a single monkey, the topographic patterns in area reorganization differed between area 3b and area 1. This is a surprising result, given that most of the thalamic input to both areas comes from the same place.)

With Merzenich we find a completely different sort of experiment than with Perachio. Merzenich's research, though done on close relatives to Perachio's macaques, is focused on the somatosensory system instead of the vestibular. Moreover, it is concerned with a single modality instead of the interaction between two, and it is concerned with how other inputs are represented in cortex as the periphery is damaged, instead of how a brain with damaged receptors continues to perform the same task. Nevertheless, and despite all these important differences, Perachio's principle that sensory systems overlap is born out by Merzenich's data. The slogan applies in both domains, though the data are otherwise wildly removed from one another.

This pattern of connecting general principles to specific experiments continues to ramify. The work of Merzenich et al. connects with functional magnetic resonance imaging studies of human somatosensory cortex, human behavioral studies of phantom limb patients, and other animal studies, which all are also consistent with the claim that the brain maintains overlapping dynamic representations (Wall and Egger 1971; Dostrovsky et al. 1976; Merrill and Wall 1978; Metzler and Marks 1979; Kelahan et al. 1981; Ramachandran 1994; Sanes et al. 1995). Though the experiments and resultant experimental data differ radically from one to the other, they all still fit under Perachio's guiding rubric.

At the end of the day, we have a set of related theoretical principles that jointly make up a general theoretical framework. And these principles are held together by the detailed data coming out of a wide variety of animal studies. In this brief chapter, we began our investigative journey with single-cell recordings in the periphery of the gerbil CNS and are ending with brain images of human somatosensory cortex. (And we have only described a tiny portion of the relevant data.) But the interpretive pattern remains clear. Neuroscience continually moves between two different ways of understanding the nervous system, first in broad and sweeping strokes and second by being submerged in the minutiae.

General theoretical principles arise out of and then feed back into particular animal experiments performed on different animal models. Because physiology differs across species, specific experimental protocols are appropriate only for specific models. Sometimes the data arising out of the different animal models or different experimental procedures overlap, but largely they do not. (Better: it is not clear whether they do.) Hence, sometimes the detailed conclusions are consistent, but sometimes—a lot of the time—they are not.

Neuroscientists weave a story through their animal models and experimental protocols united by a common guiding theoretical thread. They both find commonalties and define differences. And this entire exercise, taken together, fashions the theoretical structure of neuroscience.

NOTE

We owe thanks to Bob Richardson, who read an earlier version of this chapter and made several insightful comments regarding clarity and truth. The audience at the Pittsburgh-Konstanz conference was lively and critical; we could not have asked for a better exchange. Our thanks to them for helping us to relate our ideas to what others are doing in philosophy of science and for prodding us to think more about what biological mechanisms really are.

REFERENCES

Baarsma, E. A., and H. Collewjin. 1975. "Changes in Compensatory Eye Movements after Unilateral Labyrinthectomy in the Rabbit." *Archives of Otorhinolaryngology* 211: 219–30.

Barinaga, M. 1995. "Remapping the Motor Cortex." *Science* 268: 1696–98.

Berthoz, A. 1988. "The Role of Gaze Compensation of Vestibular Dysfunction: The Gaze Substitution Hypothesis." *Progress in Brain Research* 76: 411–20.

Clark, S. A., T. Allard, W. M. Jenkins, and M. M. Merzenich. 1988. "Receptive Fields in the Body-Surface Map in Adult Cortex Defined by Temporally Correlated Inputs." *Nature* 332: 444–45.

Dostrovsky, J. O., J. Millar, and P. D. Wall. 1976. "The Immediate Shift of Afferent Drive of Dorsal Column Nucleus Cells Following Deafferentation: A Comparison of Acute and Chronic Deafferentation in Gracile Nucleus and Spinal Cord." *Experimental Neurology* 52: 480–95.

Fisch, U. 1973. "The Vestibular Response Following Unilateral Vestibular Compensation." *Acta Oto-laryngologica* 76: 229–38.

Flohr, H., J. Bienfold, W. Abeln, and I. Macskovics. 1981. "Concepts of Vestibular Compensation." In H. Flohr and W. Precht, eds., *Lesion-Induced Neuronal Plasticity in Sensorimotor Systems*. Berlin: Springer-Verlag, 153–72.

Galiana, H. L, H. Flohr, and G. M. Jones. 1984. "A Re-evaluation of Intervestibular Nuclear Coupling: Its Role in Vestibular Compensation." *Journal of Neurophysiology* 51: 258–75.

Goldberger, M. E. 1980. "Motor Recovery after Lesions." *Trends in Neuroscience* 3: 288–91.

Hardcastle, V. G. 1995. *How to Build a Theory in Cognitive Science*. Albany: State University of New York Press.

Kelahan, A. M., R. H. Ray, L. V. Carson, C. E. Massey, and G. S. Doetsch. 1981. "Functional Reorganization of Adult Raccoon Somato-sensory Cerebral Cortex following Neonatal Digit Amputation." *Brain Research* 223: 152–59.

Kjerulf, T. D., J. T. O'Neal, W. H. Calvin, J. D. Loeser, and L. E. Westrum. 1973. "Deafferentation Effects in Lateral Cuneate Nucleus of the Cat: Correlation of Structural Changes with Firing Pattern Changes." *Experimental Neurology* 39: 86–102.

Loeser, J. D., A. A. Ward, Jr., and L. E. White, Jr. 1968. "Chronic Deafferentation of Human Spinal Cord Neurons." *Journal of Neurosurgery* 29: 48–50.

Merrill, E. G., and P. D. Wall. 1978. "Plasticity of Connection in the Adult Nervous System." In C. W. Cotman, ed., *Neuronal Plasticity.* New York: Raven Press, 94–111.

Merzenich, M. M, J. H. Kaas, M. Sur, R. J. Nelson, and D. J. Felleman. 1983. "Progression of Change Following Median Nerve Section in the Cortical Representation of the Hand in Areas 3b and 1 in Adult Owl and Squirrel Monkeys." *Neuroscience* 10: 639–65.

Metzler, J., and P. S. Marks. 1979. "Functional Changes in Cat Somatic Sensory-Motor Cortex During Short-Term Reversible Epidermal Blocks." *Brain Research* 177: 379–83.

Miles, F. A., and S. G. Lisberger. 1981. "Plasticity in the Vestibulo-ocular Reflex: A New Hypothesis." *Annual Review of Neuroscience* 4: 273–99.

Newlands, S. D., and A. A. Perachio. 1990a. "Compensation of Horizontal Canal Related Activity in the Medial Vestibular Nucleus Following Unilateral Labyrinth Ablation in the Decerebrate Gerbil. I. Type I Neurons." *Experimental Brain Research* 82: 359–72.

———. 1990b. "Compensation of Horizontal Canal Related Activity in the Medial Vestibular Nucleus Following Unilateral Labyrinth Ablation in the Decerebrate Gerbil. II. Type II Neurons." *Experimental Brain Research* 82: 373–84.

Ramachandran, V. S. 1994. "Phantom Limbs, Neglect Syndromes, Repressed Memories, and Freudian Psychology." *International Review of Neurobiology* 37: 291–372.

Sanes, J. N., J. P. Donoghue, V. Thangaraj, R. R. Edelman, and S. Warach. 1995. "Shared Neural Substrates Controlling Hand Movements in Human Motor Cortex." *Science* 268: 1775–77.

Schaefer, K.-P., and D. L. Meyer. 1973. "Compensatory Mechanisms Following Labyrinthine Lesions in the Guinea Pig: A Simple Model of Learning." In H. P. Zippel, ed., *Memory and Transfer of Information.* New York: Plenum Press, 203–32.

———. 1974. "Compensation of Vestibular Lesions." In H. H. Kornhuber, ed., *Handbook of Sensory Physiology,* vol. 2. New York: Plenum Press, 462–90.

Schaffner, K. F. 1980. "Theory Structure in the Biomedical Sciences." *Journal of Medicine and Philosophy* 5: 57–97.

———. 1993. *Discovery and Explanation in Biology and Medicine.* Chicago: University of Chicago Press.

Sirkin, D. W., W. Precht, and J. H. Courjon. 1984. "Initial, Rapid Phase of Recovery from Unilateral Vestibular Lesion in Rat Not Dependent on Survival of Central Portion of Vestibular Nerve." *Brain Research* 302: 245–56.

Smith, P. F., and I. S. Curthoys. 1989. "Mechanisms of Recovery Following a Unilateral Labyrinthectomy: A Review." *Brain Research Reviews* 14: 155–80.

Suppe, F. 1989. *The Semantic Conception of Theories and Scientific Realism.* Chicago: University of Illinois Press.

Wall, P. D., and M. D. Egger. 1971. "Formation of New Connections in Adult Rat Brains after Partial Deafferentation." *Nature New Biology* 232: 542–45.

3

On the Nature of Explanation in the Neurosciences

Antti Revonsuo

Department of Philosophy, Center for Cognitive Neuroscience, University of Turku, Turku, Finland

The nature of explanation in neuroscience has not been frequently discussed in the philosophy of science. Although there is an approach known as neurophilosophy (Churchland 1986), it has not resulted in a critical discussion of the epistemological, methodological, or explanatory problems connected with neuroscience; instead neurophilosophy is regarded merely as an expression of an eliminativist-reductionist program in the philosophy of mind. Thus it is fair to say that there is currently no philosophy of neuroscience, at least not in the same sense as there is a philosophy of physics or a philosophy of biology (Gold and Stoljar 1999).

In this chapter my aim is to expose and examine some fundamental explanatory problems in neuroscience. I begin the discussion by analyzing what "explanation" typically amounts to in empirical basic neuroscience (i.e., neuroanatomy and neurophysiology). The view that emerges is that basic neuroscience employs *mechanistic* explanations across multiple levels of description. Thus basic neuroscience concentrates on describing the *proximate* causation of neural mechanisms, but explanation in terms of universal laws or ultimate evolutionary causes is not among the main aims of basic neuroscience.

After this brief description of the principal explanatory strategy in basic neuroscience I expose some of the most difficult problems it implies. First, a mechanistic explanatory strategy includes certain assumptions concerning the systems that are to be explained. A central assumption is that the system is composed of functionally more or less autonomous or isolable parts, and that we can have sufficient epistemic access to such parts through our research instruments and experimental methods. However, these assumptions may not necessarily hold when it comes to the study of the central nervous system. Certain problems in current neuroscience suggest that, when applied to the brain, the mechanistic explanatory strategy might not be appropriate. Whether these obstacles will prove to be insurmountable and permanent or merely a passing inconvenience remains to be seen.

The second explanatory problem arises from the fact that not all neuroscience is thoroughly biological. A recently founded and rapidly growing branch entitled cognitive neuroscience attempts to integrate levels of psychological or cognitive description with those of basic neuroscience. Cognitive neuroscience has even gone so far as to declare itself "the biology of the mind" (Gazzaniga et al. 1998). Such a move obviously makes neuroscience not just any branch of the biological sciences, but the very branch that is supposed to explain psychological phenomena by referring to the neural levels of description. That is a tall order, and it implies severe explanatory problems beyond the scope of purely biological neuroscience. Two major questions emerge: How should the relationship between psychological and biological explanations be construed in cognitive neuroscience? If cognitive neuroscience is the combination of cognitive science and neuroscience—the former regards psychology as explanatorily autonomous, the latter regards all higher-level phenomena as mechanistically explainable by referring to lower-level biological and biochemical phenomena—then there is a clear conflict of explanatory strategies and assumptions built into the ingredients of cognitive neuroscience.

In the first section, I concentrate on explanation in basic neuroscience and analyze its problems. In the second, the cognitive levels of description are related to neuroscience, which leads to cognitive neuroscience, and the special explanatory problems connected with it are analyzed.

Explanation in Basic Neuroscience

Basic Neuroscience and Causal-Mechanical Explanation

In current philosophy of science there are two different views of scientific explanation. The unification approach is a descendant of the

Hempelian view, and therefore it seeks laws and principles that are as universal as possible and provide the basis of a systematized, coherent overall picture of the world. By contrast, the causal-mechanical view sees explanatory knowledge as an understanding of the hidden mechanisms by which nature works. Although these views are clearly dissimilar, they can nevertheless be taken as compatible with each other (Salmon 1989).

Which one of these views could be accommodated with explanation in the biological sciences in general and basic neuroscience in particular? Let us first briefly consider the structure of biological knowledge. Different subdisciplines in biology can be classified according to the questions they are trying to answer (Mayr 1997). All branches of biology start with meticulous *description of the relevant phenomena,* that is, by trying to answer the "What?" questions. Any new level of description in biology is created in that way. But when it comes to theoretical explanation of what has been described, there are two distinct questions to which a certain subdiscipline can provide answers. All aspects of physiology and functional biology try to answer the "How?" question. They seek an understanding of *the immediate causal mechanisms that directly bring about the phenomenon in question* (biologists often call this the "proximate causation" of the phenomenon). By contrast, evolutionary biology is concerned with "Why?" questions; an understanding of why the phenomenon exists at all, and of why it became what it is. This requires a *historical explanation in terms of natural selection and evolutionary processes* ("ultimate causation"). As Ernst Mayr (1997, 118) insists, "no biological phenomenon is fully explained until both proximate and ultimate causations are illuminated."

When I refer to basic neuroscience (as contrasted with cognitive neuroscience), what I primarily have in mind is neuroanatomy and neurophysiology at various levels of description, from the molecular level up to the level of the nervous system as a whole. Thus basic neuroscience in this sense is concerned with the "What?" and the "How?" questions, but not necessarily with the evolutionary "Why?" questions. Accordingly, it constructs accounts of the immediate or "proximate" causation of neurobiological phenomena, that is, accounts of the underlying micro-level mechanisms that constitute the phenomena to be explained. Explanation in basic neuroscience is a prime example of causal-mechanical explanation rather than explanation in terms of universal laws and principles.[1]

Additional support for this view of the nature of biological explanation comes from Bechtel (1994), who argues that biological knowledge is not primarily represented in universal laws or linguistic structures. Biologists typically first identify an interesting system at one level of organization in nature and then try to figure out what the components of this system are, how they interact, and how they produce the effects that can be observed at the level of the whole system. When they go about this task, they try to take the system apart or visualize it better with the help of various research instruments in order to figure out what the components and microstructures of the system are like. From these data biologists attempt to build an idealized model of the system, the purpose of which is to show the general structure and function of the system. The model may be only partially (if at all) clothed in linguistic representations; instead, all kinds of diagrams and figures can often best depict the component structures of, and their mutual interactions within, the biological system in question (Bechtel and Richardson 1992; Bechtel 1994).

This view of explanation is well in accordance with the actual structure of biological knowledge and the actual practice of biological research. Our understanding of a living organism forms a hierarchy of levels of description from biochemistry and molecular biology to cytology, histology, and the macroanatomy and physiology of the whole organism. These levels of description are thought to correspond to ontological levels of organization in nature, each level made up of different kinds of basic entities and their interactions, with the description and explanation of phenomena at each level requiring different vocabularies and models. Bechtel (1994) calls this framework *mechanistic explanation*.

This much seems clear: basic neuroscience, along with all other branches of anatomy and physiology, employs causal-mechanical or mechanistic explanations. It attempts to produce models of neural systems at several different levels of description: synapses and synaptic transmission, single-neuron morphology and electrophysiology, neural connectivity and organization in sensory systems and the central nervous system, cytoarchitectonics of the cerebral cortex, macroanatomy of the brain, and so forth. The role of visualization clearly is important: for example, in brain mapping, our understanding of the brain at several different levels is represented in the form of three-dimensional visual models (Toga and Mazziotta 1996). In basic neuroscience, explanation takes place when a phenomenon residing at a higher level of organization (say, an action potential fired by a single neuron) is explained by an idealized model showing how the phenomenon was brought about by the causal interactions of the parts of the system residing at lower levels of organization (e.g., membranes, membrane potentials, selective permeability, transmembrane proteins, ion gradients, the transient opening of ion channels).

Challenges for Mechanical Explanation in Basic Neuroscience

In order to employ the mechanistic explanatory strategy, we should first figure out what the different levels of organization in the nervous system are. During this century basic neuroscience has been dominated by the study of the relatively low levels of organization, concentrating especially on the level of the single neuron. Therefore, our understanding of how neurons are organized to form functional wholes at higher levels of organization in the central nervous system is still relatively poor.

The fundamental question for the mechanistic program is *how to decompose the central nervous system into functionally meaningful parts*. A part found within a system causally interacts with other parts of about the same magnitude, and a cluster of causally interacting parts make up a level of organization in the system (Bechtel 1994). Discovering the causally interacting parts of the system thus reveals the levels of organization in the system and can lead to the construction of the corresponding levels of description and explanation: a mechanistic model of the system.

However, to employ the mechanistic program in practice is not quite that simple. There are at least two different problems for the program of mechanistic explanation in basic neuroscience. The first is concerned with the basic theoretical assumptions of the mechanistic strategy; the second is a methodological problem concerned with our epistemic access to the levels of organization within the system. In the following sections I first describe these problems and then examine two examples showing how they surface within empirical neuroscience itself.

The Assumption of Decomposability. The general strategy in constructing mechanical explanation in biology has been the *decomposition* of the system into parts and the *localization* of subfunctions within these parts. Whether this strategy succeeds or fails in a given case, it at least provides us with a tractable strategy for attacking the explanatory problems that very complex biological systems present (Bechtel and Richardson 1993). Decomposition and localization may not be feasible if a system is not decomposable to a sufficient degree. Only systems composed of modular subsystems are decomposable; when it comes to the properties of subsystems, the causal interactions *within* the subsystems must be more important than those *between* the subsystems.

The problem is that we have no guarantee that the brain is decomposable across the board. We do know that at the relatively low, microscopic levels of organization the system is composed of single neurons and their synaptic connections, and that at the relatively coarse, macroscopic levels of description there are modular subsystems specialized for the processing of certain types of sensory information. Difficulties have emerged, however, in empirically identifying the causally relevant parts of the system, and these problems may be a sign that the brain is not decomposable at every level of organization to a sufficient degree to allow the construction of mechanistic explanations.

Dependence on Research Tools for Discovering the Component Parts. The establishment of the neuron doctrine at the beginning of this century illustrates how discoveries in basic neuroscience (as elsewhere in biology) are heavily dependent on the research instruments and methods available. At the end of the nineteenth century, Camillo Golgi invented a new method for staining neural tissue preparations, the famous Golgi stain. Use of this method of silver staining made it possible to lend individual neurons sharp contrast against their background in tissue preparations, thus for the first time allowing the clear visualization of the single neuron. Santiago Ramón y Cajal, using the Golgi stain, was the first to observe that there are small gaps between individual neurons; neural tissue can be divided into independent units, called neurons; it is not a continuous mass of tissue, as Golgi himself believed.

In current neuroscience the role of research instruments and the ability to visualize neural phenomena are at least as important as they were in the days when light microscopes and staining methods were the only available research techniques. Now neuroscientists are concerned not only with the single neuron but also with the detailed structural and functional mapping of the whole central nervous system in vivo. New brain sensing and imaging methods that until recently were unimaginable have been introduced during the past ten to twenty years and are now widely used in the research community. They are producing a wealth of data that, it is hoped, will reveal what the causally relevant parts of the system are and how they should be best described and systematized. Projects to generate detailed maps of the living human brain have been launched (Toga and Mazziotta 1996). At the moment the main efforts are concentrated on the description and systematization of data and the utilization of these data for clinical purposes. No radically new theoretical principles, comparable to the neuron doctrine, have emerged from this enterprise as yet.

The Problem of Identifying the Parts of the System

The problems in identifying the functionally relevant parts of the system are clearly manifested in the disagreements over what the basic principles of neural coding are and how the basic divisions of the visual cortex should be delineated.

How Do Neurons Code Information? A theory of neural coding should describe those functionally relevant parts and mechanisms of the system that carry sensory information and transform it into perceptual and motor information in the brain and, finally, into observable behavioral outputs. At least three competing solutions to the problem of neural coding have been proposed: the grandmother cell theory, the cell assembly theory, and the temporal coding theory. According to the grandmother cell theory, there is one neuron coding every perceptual element, and these neurons project, hierarchically and convergently, to higher-order neurons, which then code parts of objects and finally whole objects. When a cell high in the hierarchy increases its firing rate, it signifies the presence of the coded object in perception. According to the cell assembly theory, a perceptual object is coded as an activation pattern in a large network of neurons, where each cell is coding a simple feature of the object. Many different objects can be represented by the same network as different combinations of features. The representations are rate coded: changes in the information a cell represents imply changes in the rate of firing. According to the temporal coding theory, rate coding is not sufficient to explain the coding of perceptual information. Our perception can represent more than one object at a time; therefore there must be some mechanism that distinguishes one activation pattern from another in the network. A single unified pattern is temporally coded, that is, all the cells belonging to one pattern fire together synchronously; if there are other unified patterns active at the same time, representing other perceptual objects, their constitutive cells must establish a different rhythm of synchronous firing (Fotheringhame and Young 1997; Gazzaniga et al. 1998).

Although the grandmother cell theory has been almost entirely discarded, there is still little agreement about the level of organization at which informationally and functionally important neural activity resides. There is a growing body of experimental evidence for the role of neural synchronization in perceptual processing (Engel et al. 1999a; Gray 1999; Sauvé 1999), but disagreements abound over how this evidence should be interpreted (Engel et al. 1999b; Vanni 1999).

These disagreements are a symptom of rather difficult theoretical and methodological problems in identifying the functionally relevant parts of the system and, therefore, the actual levels of organization in the brain. The question is: *How does one separate the actual causally efficient processes from all the rest of the phenomena that can be observed or measured at some level of organization in the brain?* Bullock (1997) discusses this problem and divides the current views on how the information-bearing elements in our nervous system communicate into the *unit window view* and the *population window view.* According to the unit window view, "Neural communication consists of successions of nerve impulses in neurons, encoding messages in their intervals, decoding at axonal terminals into an analog dose of transmitter that restarts the cycle in the next cell. A principal problem in explaining higher functions is seen as the adequacy of sampling of units" (Bullock 1997, 1).

The contrasting population window view denies that neural communication could be exhaustively described in this manner. Organized cell assemblies have, in addition to single-unit spike activity, other causally relevant properties that appear at a level of organization higher than the single neuron. Major features of the dynamics of organized cell assemblies include "their nonspike, more slowly fluctuating potentials, their changing degrees of population synchrony, and their rhythms and large-scale patterns" (Bullock 1997, 1).

It is as yet completely unclear which of these electrophysiological phenomena realize causal-functional roles in the overall functional organization of the system, and which are mere nonfunctional signs of activity. In other words, *we do not know which of these phenomena we should try to measure and model* when we are building a mechanical explanation, for *we do not know which (if any) of these phenomena constitute actual levels of organization in the brain.* Bullock (1997) admits that our ignorance of the kinds of operations going on in neural assemblies is so profound that we cannot anticipate what measures are the most relevant ones to use. He argues that major new levels of biological organization will have to be uncovered in order to understand what is going on at the multiple levels of electrophysiological activity in the brain.

The problem of neural coding is thus a manifestation of the more fundamental problem associated with the mechanistic explanatory strategy: the identification of the system's functional parts and the levels of organization that they constitute. It seems fair to conclude that currently we do not really know the fundamental principles of functional organization in the brain. Is this merely the current state of affairs or a symptom of the fact that the brain may not conform to the assumptions that mechanistic explanation entails?

How Many Visual Cortical Areas? The other issue in connection with which a similar problem about determining the functionally relevant parts of the system has lately surfaced is the identification of different visual cortical areas. Everybody in the field agrees where the primary visual cortex (V1) is situated and how it should be demarcated from other brain areas. This is because there are multiple converging criteria that all point to the fact that this is an anatomically and functionally distinct area and therefore a genuine, relatively independent part of the system. Furthermore, it is generally accepted that there are multiple visual areas (Zeki and Bartels 1999). However, there is no general agreement either on the number of distinct visual cortical areas or on their individual functional specializations. The same criteria (e.g., retinotopy, histology, connectivity, global functional properties) that can be unambiguously used to demarcate early visual areas from each other become more controversial at later stages in the visual system. There is a core group of visual areas (e.g., V1, V2,

V3/VP, MT) for which the different identification criteria converge with each other reasonably well. Different research groups, independent of each other, have been able to identify these areas. However, in higher visual areas the same set of criteria cannot be consistently applied. Higher areas are not retinotopic, and the areas indicated by functional criteria do not correspond well with those defined by the classical histological areas of Brodmann (Tootell et al. 1998). Therefore retinotopy, anatomy, and function diverge from each other in these areas. (For recent disagreements about identifying the visual areas V4, V8, V4V, and V4v, see note 1 in Zeki and Bartels 1999.)

The fundamental problem behind the difficulties in constructing a universal map of the visual cortex is that everything in the extrastriate cortex is densely connected with almost everything else, and *it is very difficult to say what would constitute a sufficient criterion—or set of criteria—for demarcating an area of cortex as separate from the surrounding parts* (i.e., as a genuine subsystem). These heavy interconnections between the proposed different cortical areas may be an indication that these cortical subsystems are not modular to a sufficient degree to allow decomposition and localization. If that is the case, (some of) the identified subsystems may not exist anywhere outside the brain researchers' maps.

In cognitive neuroscience, there seems to be the assumption that functional brain imaging coupled with specified cognitive tasks can reveal the functionally relevant parts of the brain. Functional criteria are thus regarded as sufficient when identifying the parts of the system. Recently, new visual cortical areas have been identified on the basis of exactly such criteria, for example the fusiform face area and the parahippocampal place area (Kanwisher et al. 1997a, 1999; Tong et al. 1998), which are selectively activated by visually perceived faces and places, respectively.

Resorting to functional criteria does not necessarily solve the problem. Functional brain imaging involves a host of methodological difficulties of its own. In this context the most critical of them is that the cognitive models and subtraction methodology (the methodology for subtracting the brain activation measured in the control condition from that in the experimental condition) used in the experimental designs simply *assume* modularity and functional decomposability of the system (at the cognitive level). Even if this assumption is wrong, the subtraction images practically always show *some* differences between two cognitive tasks. A subtraction image typically shows that certain parts of the brain were more active in one condition than in another. Without the assumption of modularity, already inherent in the cognitive models used to design the task conditions, the subtractions themselves cannot establish that the activated regions constitute real subsystems or genuine parts of the system (Van Orden and Paap 1997).

Conclusion

Explanation in basic neuroscience can primarily be thought of as causal-mechanical explanation. This contrasts with what Salmon (1989) calls the unification approach, according to which explanation is based on universal principles and laws. The unification approach is probably more easily compatible with explanation in physics than with explanation in biology and neuroscience.[2]

The mechanistic program in basic neuroscience is, however, not without its problems. There are severe difficulties in identifying the functionally important parts of the system and in defining the proper levels of organization and explanation. These difficulties emerge both at relatively low levels of organization, when trying to move from the bottom up, from single neurons to organized cell assemblies, and at relatively high levels of organization, when trying to move from the top down and figure out how the entire cerebral cortex is divided into functional parts. These difficulties may, on the one hand, simply be a philosophically innocent reflection of the limited means by which we can access the system with the research instruments currently available. On the other hand, they may unfortunately be a sign of a fatal flaw in the assumptions of the mechanistic research strategy employed. If that is the case, the mechanistic strategy will sooner or later become permanently unproductive when applied to the study of the brain.

Explanation in Cognitive Neuroscience

The Challenge of Integrating Psychological and Neural Levels

Cognitive neuroscience is a relatively recent branch of neuroscience. It emerged during the 1980s, with the establishment of the *Journal of Cognitive Neuroscience* in 1989. It grew rapidly during the 1990s; landmark publications defining the scope of the field are the 1,500-page volume *The Cognitive Neurosciences* (Gazzaniga 1995) and the textbook *Cognitive Neuroscience: The Biology of the Mind* (Gazzaniga et al. 1998).

While basic neuroscience applies causal-mechanical explanations in neuroanatomy and neurophysiology, it seems that cognitive neuroscience attempts to extend the strategies of mechanistic biological explanation all the way up to psychological phenomena. This goal is obvious if we see how some of the major advocates of cognitive neuroscience characterize their field:

Cognitive neuroscience is an attempt to understand how cognition arises from brain processes; the focus is on the brain, as the term "neuroscience" implies. We don't want to separate the theory of information processing from the theory of the brain as a physical mechanism. . . . A complete cognitive neuroscience theory would specify . . . how each process is instantiated in the brain, and how brain circuits produce the input/output mappings accomplished by each process. . . . Given these goals, it seems clear that cognitive neuroscience must move closer to neurobiology. But it will simply not become neurobiology. (Kosslyn 1997, 159–60)

At some point in the future, cognitive neuroscience will be able to describe the algorithms that drive structural neural elements into the physiological activity that results in perception, cognition, and perhaps even consciousness. . . . The future of the field, however, is working toward a science that truly relates brain and cognition in a mechanistic way. . . . The science built up to understand how the brain enables the mind has come to be called cognitive neuroscience. (Gazzaniga 1995, xiii)

Cognitive neuroscience sees psychological levels (conceptualized as, e.g., "cognition," "information processing," "representation," "computation") as the higher levels of description, to be explained by referring to the neural and neurocomputational mechanisms residing at the lower levels. In this view, psychological phenomena are not explanatorily autonomous, but neither are they eliminable—just like cytology is neither eliminable nor autonomous in relation to biochemistry and molecular biology. Psychological properties are regarded as residing at a level of organization higher than neural properties, but nevertheless as being *micro-based properties* (Kim 1997) essentially in the same sense as are other special-science properties.

The cognitive neuroscientific view of explanation is radically different from the standard computationalist and representationalist views in cognitive science, which hold that psychology resides at an explanatorily autonomous level. The original idea, embedded in the very foundations of cognitive science and functionalistic philosophy, was to stay clear of biology and neuroscience. Folk psychology (propositional attitudes) and cognitive models of human information processing were at most decomposable to a basic computational description (e.g., "the language of thought"), but the computational or algorithmic level was "multiply realizable"—an abstract functional description, the computer program of the mind, which could not be reduced to the materials of the device (brain or computer) that happened to realize it. Thus cognitive science preserved both the autonomy of psychological explanation and token materialism, and for a while almost everybody was happy with these solutions. Even the most recent attempts to get rid of traditional computationalism by defining cognitive science as the science of dynamical systems (van Gelder 1998) are still committed to functionalism and "provide models that stand quite independently of implementation details" (van Gelder 1998, 660), that is, the models do not refer to any neural levels of description when describing and explaining human cognition.

The crucial difference between the cognitive science and the cognitive neuroscience view of psychological properties is that the former defines psychological properties as higher-*order* (i.e., functional) properties, whereas the latter regards them as macroproperties residing at higher *levels of organization* in the micro-macro hierarchy. *Higher-order (functional, computational) properties,* however, reside at the *same* level of organization as their realizing (physical-neural) properties, and they have *no causal powers of their own;* all the causal efficacy is possessed by their first-order realizing properties. In contrast, *macroproperties* at higher levels of organization *do have causal powers that go beyond the causal powers of their microconstituents* (Kim 1997). In accordance with the view that the explanation of a macroproperty requires the description of its lower-level microconstituents, cognitive neuroscience holds that *reference to at least some neural levels of description is necessary in the explanation of psychological phenomena:* "a science that truly relates brain and cognition in a mechanistic way" (Gazzaniga 1995, xiii). "We don't want to separate the theory of information processing from the theory of the brain as a physical mechanism" (Kosslyn 1997, 159). By contrast, cognitive science and functionalism hold that *psychological explanations are completely independent of the neural levels of description.*

There is some confusion between these two views of psychological properties in cognitive neuroscience. Although functional properties are second-order properties without causal powers of their own and defined without specifying the physical-neural properties that realize them, cognitive neuroscientists seem to treat functional properties as if they were causally effective micro-based properties at a higher level of organization than neural properties. That is, they seem to believe that an explanation of cognitive phenomena is given by first specifying at a higher level of organization the "computations," or "input-output mappings," of each neural circuit and system, and that these "algorithms" consequently "drive structural neural elements into physiological activity" (Gazzaniga 1995, xiii). These formulations strongly suggest that *cognitive neuroscientists confuse the second-order functional properties of neural systems with their higher-level macroproperties.* They seem to believe that there is a cognitive or mental level of organization in the brain where "algorithms" and other functionally defined entities establish the causally efficient properties of the system.

The problem is that the functionalistic notions do not name any natural phenomena or genuine micro-based higher-level properties, but rather formal descriptions of input-output mappings in neural systems. Functional descriptions do not define any concrete levels of organization in the brain; they are instead abstract notions that can be used to describe any natural phenomena at any level of organization, if we so choose. The higher-order descriptions are multiply realizable, but also observer-relative, not intrinsic biological levels of organization in the system itself (Searle 1992; Revonsuo 1994). Since they do not describe the actual causally effective mechanisms in the system, they cannot constitute the causal-mechanical explanation of the system. Uncritically adopting such notions in cognitive neuroscience leads to a confusion between causal-mechanical explanation, as employed in basic neuroscience, and functionalistic explanation, as practiced in traditional cognitive science.

The functional descriptions have no independent causal powers in the system. Unlike the first-order physical-neural properties, functional properties cannot cause anything in the system; they are not concrete entities existing within the system or interacting with the other parts of the system. Nevertheless, Gazzaniga (1995, xiii) says that algorithms "drive structural neural elements into physiological activity," as if an algorithm was a real entity or process within the neurobiological system, with causal powers to "drive" the structural elements into this or that activity. Thus an abstract, observer-relative functionalistic description is taken to be a concrete thing with causal powers within the system.

This mistake indicates that empirical cognitive neuroscientists have not noticed the fact that the explanatory strategies in basic neuroscience and cognitive science are very different; therefore in cognitive neuroscience one cannot adopt both of them simultaneously without committing a fundamental philosophical error. Causal-mechanical explanation, as in basic neuroscience, and functionalistic explanation, as in cognitive science, are incompatible.

This is not to say that the description of input-output mappings or functions would be useless or banned in basic neuroscience. By contrast, neuroscientists do describe the activities of neural entities as input-output functions. These functions characterize the activity of some entity at an abstract level without reference to its context or its internal structure (see Craver 2001). Defining the complex input-output function of some type of neural activity can be very useful when the scientist wants to describe the activity at an abstract level, without characterizing the details of the mechanisms that instantiate the function. The input-output function is also important for providing constraints or framing the constitutive explanations that do describe the complexities of the internal mechanisms of the system (Craver 2001).

However, the crucial difference between cognitive science and basic neuroscience explanations is that for the latter input-output functions are only one ingredient of the complete description and explanation of a neural mechanism, whereas for the former they are the ultimate explanation. Neuroscientists look at the lower levels of organization in order to find the internal causal mechanisms that actually explain why a certain input-output function is realized by the system. Cognitive scientists, however, believe that cognitive explanations should not mention any facts about the implementation of the function; the explanation comes to an end at the cognitive level when all the algorithms involved in the input-output function have been described. The details of the physical implementation of the function add nothing to the explanation; they are explanatorily irrelevant.

Cognitive neuroscience must somehow reconcile these opposing explanatory strategies. If the mechanistic explanatory strategy is to be fully applied in the explanation of mental and cognitive phenomena, then it seems that the functionalistic commitments to the autonomy of psychology must give way. Cognitive neuroscience should construct genuine cross-level explanations that show how the internal neural mechanisms at lower levels of organization give rise to mental phenomena at higher levels of organization. The descriptions of the input-output functions at each level only constitute partial intralevel descriptions, not full mechanistic explanations, of the phenomena.

The Methods of Cognitive Neuroscience: Imaging the Mind?

The new methods for sensing and imaging the brain have been hailed by some cognitive neuroscientists as methods by which we can "see the mind" (Posner 1993) or produce "images of mind" (Posner and Raichle 1994). There are, however, several methodological and epistemological problems with such claims.

Here I present only a superficial overview of these issues (for a more detailed critique, see Van Orden and Paap 1997). The basic idea in cognitive neuroscience is to take a cognitive model of a certain mental process (say, object recognition or speech comprehension) and then design an experiment to figure out when and where the brain performs the different information processing subtasks involved. The cognitive models used include several contestable assumptions about mental functioning in the brain: cognitive processing is assumed to be modular; the component modules are assumed to be independent of each other; causal connections between the modules are assumed to be only linear, feed-forward links; the corresponding functional modules are assumed to exist anatomically in the brain; it is assumed that any one component can be added or subtracted by turning it on or off, without affecting the components adjacent to it.

Experiments in cognitive neuroscience are typically designed by relying on such models. The researcher identifies one processing stage of interest (say, shape perception in object recognition or phonological processing in language comprehension), identifies the corresponding module in the cognitive model, and then designs a control condition and an experimental task in which the involvement of the critical component is manipulated (e.g., by changing the stimuli or the task) but everything else is kept constant. Then the subjects perform the experimental task and the control condition during an electroencephalographic (EEG), magnetoencephalographic (MEG), positron emission tomographic (PET), or functional magnetic resonance imaging (fMRI) measurement, and the differences in brain activity between the conditions are assumed to reflect the contribution of the critical component of interest, revealing its temporal latency, anatomical location, or both.

There are a huge number of assumptions behind such an experimental paradigm, any one of which might be wrong. Our cognitive models may be erroneous; there is no guarantee that we have discovered the actual components of cognition. In fact, Van Orden and Paap (1997) go so far as to say that no single cognitive component has yet been discovered for which there is general agreement among investigators. Furthermore, the cognitive models seem to be overly simplistic if compared with the actual organization of neural connectivity in the brain. For example, in the visual cortex, about thirty to forty different visual areas have been identified, and these areas are heavily interconnected. Simple linear cognitive models are unlikely to capture what really is going on in the brain when sensory information is processed.

Functional brain imaging methods such as PET and fMRI require the use of relative measures: what is measured is the relative difference between two conditions rather than the absolute activation during a single condition. This sort of paired-subtraction method must assume that during the experiment the subjects activate only information relevant to the performance of the task and implement as few operations as necessary for its completion: nothing but the necessary components must be activated in both the control and experimental conditions. However, we have no guarantee that this is actually the case. If a task is too easy, the subjects tend to start thinking about other things, which may cause the uncontrolled activation of several brain regions. If a task is incorrectly designed or the model on which it is based is incorrect, there may be differences, completely unanticipated by the researchers, between the control and the experimental task. However, since the control and the experimental tasks are almost inevitably different in *some* ways, the subtraction methodology is as good as guaranteed to show some brain areas "light up" in the subtraction image. Thus even an incorrectly designed task will most likely give a neat "result."

The interpretation of the differences in brain activation is extremely difficult. The brain images only reveal that one region is more active than another. This can be interpreted as showing that the activated regions are *somehow* involved in the task, but they show nothing about *how* the task is performed or *what* the activated area is doing.

The simplest interpretation is that the area is processing exactly the kind of information that we tried to manipulate between the conditions. But other interpretations are feasible: maybe the activation reflects the general *effort* put into the task, not the particular processing stage we assume it to reflect; maybe the brain area is actually an *inhibitory* area that stops some other areas from disturbing the task, but the area itself does not otherwise contribute to carrying out the task.

Then there are worries about the observed events themselves: what exactly are we measuring or observing with EEG, MEG, PET, or fMRI? Where do the signals come from, and why should we believe that they have anything to do with mental events, or that they reflect any functionally relevant levels of organization in the brain? Noninvasive scalp-recorded EEG and MEG actually detect fast changes in the gross synchronous activity of millions of synapses in selected areas near or on the cortical surface. If the current sources are too small (involving only a few thousand postsynaptic potentials) or incoherently oriented or asynchronously activated or too deep below the cortical surface, they will not generate any signals detectable with EEG or MEG. Furthermore, although the temporal resolution of EEG and MEG is good (one millisecond), we can never directly observe or even indirectly calculate the location of the original electrophysiological processes in the cortex. The "inverse problem" makes it impossible to calculate the location of the intracranial source of an electrical or magnetic field: there are infinitely many mathematically possible solutions. Only an educated guess—a source model—can be constructed on the basis of the indirect observations.

Thus when we evaluate the results obtained with these methods we should remember that, at best, we see gross spatial averages of fast changes in the synchronous activity of millions of synapses whose location we cannot observe but only model. Many electrophysiological processes not detectable by these methods are simultaneously going on at multiple locations and multiple levels of organization, but we have no epistemic access to them if our research instruments cannot detect them. Neither do we really know whether EEG and MEG signals reflect the activity of a certain functionally important level of organization in the brain or simply the grossly averaged humming electrical noise arising from lower levels of functional organization. Even if we assume that we are picking up signals from a functionally relevant level of organization, we still have little idea what exactly is going on in the neural population where those 10^7 synapses were coherently activated, generating the signal, and how that activity relates to activity elsewhere in the brain. How should we describe such a massively complex bioelectrical phenomenon?

In PET and fMRI it is possible to locate the measured brain activation. The problem is that neither of these methods detects signals that are directly generated by neural activity. Instead PET detects gamma rays that originated when a positron was annihilated in the brain; the positron was generated when a radioactive isotope in the tracer compound (e.g., labeled water or glucose analogue molecules) in the bloodstream decayed. The idea is that the distribution of the radioactive tracer reflects where blood flow or glucose metabolism increases in the brain relative to a control condition, and this distribution is supposed to give a general idea of "brain activation" in the experimental condition. In fMRI it is the relative amount of oxygenated versus deoxygenated blood that is detected, for these have slightly differing magnetic properties. Thus we are getting a very indirect image of neural activation, for we are directly observing only hemodynamic changes taking place in the circulatory system in the brain. Furthermore the temporal resolution of PET and fMRI is poor; the theoretical limit is about six seconds, for it takes this much time before blood flow reacts to an increase in neural activity. Last but not least, the relationship between neural activity and increased blood flow is not well understood.

In conclusion, we have every reason to be cautious in our interpretations of brain sensing and imaging data from cognitive experiments. We may have inaccurate cognitive models to begin with, and our observations of brain activity are, even at best, indirect and coarse. As a result of this, brain activation images from different studies and laboratories may diverge rather than converge. For example, the attempts to localize the phonology component have not been particularly successful (Poeppel 1996): in five different studies twenty-two different areas were highlighted. Only eight of these appeared in more than one study, and none of these regions appeared in four or all five studies. Thus these subtractions are unlikely to isolate any actual components of the system. As we pointed out in the discussion of visual cortical areas earlier in this chapter, it may be that the assumption of decomposability is generally misguided. However, in some cases at least, functional brain imaging can yield strongly converging results. To take an example, an extrastriatal visual area named the lateral occipital cortex has been identified as being involved in the processing of coherent visual shape in several independent studies using different methods (PET, fMRI, MEG) and different kinds of stimuli (Kanwisher et al. 1996, 1997b; Vanni et al. 1996; Grill-Spector et al. 1998). This convergence of results shows that in carefully designed studies the results may be interpretable. Nevertheless cognitive neuroscience methods do not provide us with "images of mind."

Cognitive Neuroscience and the Problem of Consciousness

If cognitive neuroscience is the biology of the mind, it must also be the biology of consciousness, for the possession of phenomenal consciousness is what demarcates genuinely mental systems—subjects—from nonmental zombies and other mere objects. It has been clearly accepted that consciousness belongs to the domain of cognitive neuroscience; all the major new textbooks and overviews of cognitive neuroscience contain a section devoted to consciousness (Gazzaniga 1995; Rugg 1997; Gazzaniga et al. 1998).

But how could cognitive neuroscience attack consciousness empirically? Can the same research strategies that have been used elsewhere in the biological sciences be used to address the problem of consciousness? After all, several philosophers have argued that consciousness is beyond the reach of empirical science in general and brain research in particular: even a complete description of the brain is argued not to be sufficient to enable us to understand consciousness (e.g., Chalmers 1996). The explanatory gap (Levine 1983) remains.

I believe that if we wish to understand consciousness scientifically a cognitive neuroscientific approach is the best choice. Current cognitive neuroscience, however, does not offer sufficient conceptual tools for this approach because it operates only at the cognitive levels of description on the one hand and the neural levels of description on the other. In cognitive neuroscience, if we are to take consciousness seriously, we will have to treat it as a real natural biological phenomenon that is best thought of as comprising a distinct level of organization in the brain. I call this the phenomenal level of organization. In order to capture this level systematically, what we need is a corresponding level of description and theoretical inquiry into the relations of this level to the other levels of organization in the brain.

It is obvious that currently we do not have any systematic account of the phenomenal level, and it may be doubtful whether such a description can be constructed. But we should not take the present state of affairs as a sign that our efforts are doomed from the start. The scientific study of all natural biological phenomena starts with description, and there is no way to know a priori what is going to be the best way to systematize, classify, and describe the data. Still, what we learn from the history of science is that instead of giving up we should try to construct a fruitful metaphor for the phenomenal level of organization and to identify particularly useful model systems, in order to study the phenomenon in the most revealing cases. Metaphors and model systems are widely used in biological research. Genes were depicted as units linearly arranged like beads on a string; this was a useful metaphor when the molecular composition of genes was unknown. Genetics has used model organisms, such as *Drosophila,* to control, isolate, and describe phenomena that would otherwise be difficult to study. I am suggesting that the cognitive neuroscience of consciousness should simply follow the same strategy. It should initiate the search for useful metaphors and model systems of the phenomenal level of organization in the brain.

Since I have analyzed the problems of and prospects for this kind of cognitive neuroscience approach to consciousness in more detail elsewhere (Revonsuo 1997, 2000), I will say no more about consciousness here. I merely emphasize that when we try to figure out whether or not neuroscience can (someday) explain consciousness, we should look at the ways in which complex biological phenomena are explained in general, and then figure out how the same strategies of explanation could be used in the study of consciousness in particular.

Conclusion

Basic neuroscience is an enterprise in the causal-mechanical explanation of neurobiological phenomena residing at multiple levels of organization, and it is therefore closely connected with the mechanistic research strategies dominant in the biomedical sciences. A thorough understanding of the system is at present restricted to relatively low levels of organization, and there probably are several higher levels that neuroscience has not discovered yet. This is a challenge to the mechanistic explanatory strategy, but so far it remains unclear whether the brain as a complex biological system conforms to the assumptions built into the mechanistic strategy. The brain may not be decomposable into functionally isolable parts at every level of organization. Even if it were, basic neuroscience has a rather restricted epistemic access to the multiple levels of organization in the brain. Thus it will not be an easy task to figure out which of the phenomena detected by our instruments are directly reflecting functional neural organization and which are merely inconsequential side effects doing little if any organized causal work in the system. Disagreements among neuroscientists on the nature of neural coding and on the number of visual cortical areas are revealing, but it is too early to decide whether these difficulties can be overcome without abandoning the assumptions of the mechanistic program altogether.

Cognitive neuroscience is an attempt to explain mental or cognitive levels of description in the mechanistic way. The conflicting explanatory strategies and assumptions in cognitive science and neuroscience crash at the interface of these sciences. Cognitive science is committed to functionalism and the autonomy of psychology or the total independence of psychological and neural levels of description. Cognitive neuroscience apparently attempts to adopt the best of both worlds at the same time, treating the basic abstract theoretical entities of cognitive science as if they were concrete natural phenomena entering into causal interactions with the neural levels. This confusion constitutes a serious unresolved problem for the theoretical basis of cognitive neuroscience. I suggest that the cognitivist-functionalist strategy and the autonomy of psychology should be abandoned and a full-scale mechanistic program applied to cognitive neuroscience.

The methodological and epistemological problems of cognitive neuroscience are connected to the assumptions about mental phenomena included in cognitive models: they are overly simplistic and hard to relate to the reality found in the brain. Furthermore, the signals that current brain sensing and imaging methods can pick from the brain offer but a seriously limited and blurred window on the complex neural and electrophysiological dynamics that actually exists there. Finally, consciousness presents the true challenge to cognitive neuroscience. I believe that cognitive neuroscience is the right weapon with which to attack the problem, but the attack will require that radically novel levels of description be included within cognitive neuroscience to capture and explain the reality of phenomenal consciousness.

NOTES

I am grateful to Simo Vihjanen for comments on an earlier draft. The writing of this chapter was supported by the Academy of Finland (projects 36106 and 45704).

1. This is not to say that explanation in basic neuroscience has nothing to do with universal laws and generalizations. The practice of research involves a huge number of background assumptions concerning, for example, the functioning of the research instruments and reliance on our knowledge about physical laws.

2. What about explanation in chemistry, the discipline that lies between biology and physics in the hierarchical model of the structure of science? There has until now been no philosophy of chemistry, but the situation is changing (see, e.g., Guterman 1998). Tired of what they call "physics imperialism," chemists and philosophers interested in the foundational questions of chemistry have begun to argue that molecular structure establishes an ontological level of organization that cannot be exhaustively reduced to quantum mechanics. This level can be understood with the help of idealized analogical macroscopic models (Del Re 1998). This view of explanation in chemistry approaches the causal-mechanical view and apparently tries to do away with the unification view, according to which everything follows from universal physical principles. Physics seems to remain the only science in which explanation actually might conform to the ideals of the unification approach.

REFERENCES

Bechtel, W. 1994. "Levels of Description and Explanation in Cognitive Science." *Minds and Machines* 4: 1–25.

Bechtel, W., and R. C. Richardson. 1992. "Emergent Phenomena and Complex Systems." In A. Beckermann, H. Flohr, and J. Kim, eds., *Emergence or Reduction? Essays on the Prospects of Nonreductive Physicalism*. Berlin: de Gruyter, 257–88.

——. 1993. *Discovering Complexity: Decomposition and Localization as Strategies in Scientific Research*. Princeton, N.J.: Princeton University Press.

Bullock, T. H. 1997. "Signals and Signs in the Nervous System: The Dynamic Anatomy of Electrical Activity Is Probably Information-Rich." *Proceedings of the National Academy of Sciences of the USA* 94: 1–6.

Chalmers, D. J. 1996. *The Conscious Mind*. New York: Oxford University Press.

Churchland, P. S. 1986. *Neurophilosophy*. Cambridge, Mass.: MIT Press.

Craver, C. F. 2001. "Role Functions, Mechanisms, and Hierarchy." *Philosophy of Science* (in press).

Del Re, G. 1998. "Ontological Status of Molecular Structure." *HYLE—An International Journal for the Philosophy of Chemistry* 4(2): 81–103.

Engel, A. K., P. Fries, P. König, M. Brecht, and W. Singer. 1999a. "Temporal Binding, Binocular Rivalry, and Consciousness." *Consciousness and Cognition* 8(2): 128–51.

——. 1999b. "Does Time Help to Understand Consciousness?" *Consciousness and Cognition* 8(2): 260–68.

Fotheringhame, D. K., and M. P. Young. 1997. "Neural Coding Schemes for Sensory Representation: Theoretical Proposals and Empirical Evidence." In Rugg 1997, 47–76.

Gazzaniga, M. S., ed. 1995. *The Cognitive Neurosciences*. Cambridge, Mass.: MIT Press.

Gazzaniga, M. S., R. B. Ivry, and G. R. Mangun. 1998. *Cognitive Neuroscience: The Biology of the Mind*. New York: Norton.

Gold, I., and D. Stoljar. 1999. "A Neuron Doctrine in the Philosophy of Neuroscience." *Behavioral and Brain Sciences* 22: 809–69.

Gray, C. M. 1999. "The Temporal Correlation Hypothesis of Visual Feature Integration: Still Alive and Well." *Neuron* 24: 31–47.

Grill-Spector, K., T. Kushnir, S. Edelman, Y. Itzchak, and R. Malach. 1998. "Cue-Invariant Inactivation in Object-Related Areas of the Human Occipital Lobe." *Neuron* 21: 191–202.

Guterman, L. 1998. "I React Therefore I Am." *New Scientist,* November 21, 34–37.

Kanwisher, N., M. M. Chun, J. McDermott, and P. J. Ledden. 1996. "Functional Imaging of Human Visual Recognition." *Cognitive Brain Research* 5: 55–67.

Kanwisher, N., J. McDermott, and M. M. Chun. 1997a. "The Fusiform Face Area: A Module in the Human Extrastriate Cortex Specialized for Face Perception." *Journal of Neuroscience* 17: 4302–11.

Kanwisher, N., R. P. Woods, M. Iacoboni, and J. C. Mazziotta. 1997b. "A Locus in Human Extrastriate Cortex for Visual Shape Analysis." *Journal of Cognitive Neuroscience* 9: 133–42.

Kanwisher, N., D. Stanley, and A. Harris. 1999. "The Fusiform Face Area Is Selective for Faces Not Animals." *NeuroReport* 10: 183–85.

Kim, J. 1997. "Does the Problem of Mental Causation Generalize?" *Proceedings of the Aristotelian Society* 97(3): 281–97.

Kosslyn, S. M. 1997. "Mental Imagery." In M. S. Gazzaniga, ed., *Conversations in the Cognitive Neurosciences*. Cambridge, Mass.: MIT Press, 155–74.

Levine, J. 1983. "Materialism and Qualia: The Explanatory Gap." *Pacific Philosophical Quarterly* 64: 354–61.

Mayr, E. 1997. *This Is Biology*. Cambridge, Mass.: Belknap Press/Harvard University Press.

Poeppel, D. 1996. "A Critical Review of PET Studies of Phonological Processing." *Brain and Language* 55: 322–51.

Posner, M. I. 1993. "Seeing the Mind." *Science* 262: 673–74.

Posner, M. I., and M. E. Raichle. 1994. *Images of Mind*. New York: Scientific American Library.

Revonsuo, A. 1994. "In Search of the Science of Consciousness." In A. Revonsuo and M. Kamppinen, eds., *Consciousness in Philosophy and Cognitive Neuroscience*. Hillsdale, N.J.: Lawrence Erlbaum, 249–85.

——. 1997. "How to Take Consciousness Seriously in Cognitive Neuroscience." *Communication and Cognition* 30: 185–206.

———. 2000. "Prospects for a Scientific Research Program on Consciousness." In T. Metzinger, ed., *Neural Correlates of Consciousness*. Cambridge, Mass.: MIT Press, 57–75.

Rugg, M. D., ed. 1997. *Cognitive Neuroscience*. Cambridge, Mass.: MIT Press.

Salmon, W. C. 1989. "Four Decades of Scientific Explanation." In P. Kitcher and W. C. Salmon, eds., *Scientific Explanation*. Minnesota Studies in the Philosophy of Science, vol. 13. Minneapolis: University of Minnesota Press, 3–195.

Sauvé, K. 1999. "Gamma-Band Synchronous Oscillations: Recent Evidence Regarding Their Functional Significance." *Consciousness and Cognition* 8(2): 213–24.

Searle, J. R. 1992. *The Rediscovery of the Mind*. Cambridge, Mass.: MIT Press.

Toga, A. W., and J. C. Mazziotta. 1996. *Brain Mapping*. San Diego: Academic Press.

Tong, F., K. Nakayama, J. T. Vaughan, and N. Kanwisher. 1998. "Binocular Rivalry and Visual Awareness in Human Extrastriate Cortex." *Neuron* 21: 753–59.

Tootell, R. B. H., N. K. Hadjikhani, J. D. Mendola, S. Marrett, and A. M. Dale. 1998. "From Retinotopy to Recognition: fMRI in Human Visual Cortex." *Trends in Cognitive Sciences* 2: 174–83.

Van Gelder, T. 1998. "The Dynamical Hypothesis in Cognitive Science." *Behavioral and Brain Sciences* 21: 615–65.

Vanni, S. 1999. "Neural Synchrony and Dynamic Connectivity." *Consciousness and Cognition* 8(2): 159–63.

Vanni, S., A. Revonsuo, J. Saarinen, and R. Hari. 1996. "Visual Awareness of Objects Correlates with Activity of Right Occipital Cortex." *NeuroReport* 8: 183–86.

Van Orden, G. C., and K. R. Paap. 1997. "Functional Neuroimages Fail to Discover Pieces of Mind in the Parts of the Brain." *Philosophy of Science* 64: S85–S94.

Zeki, S., and A. Bartels. 1999. "Towards a Theory of Visual Consciousness." *Consciousness and Cognition* 8(2): 225–59.

4

Mechanisms, Coherence, and the Place of Psychology

Commentary on Revonsuo

Stephan Hartmann
Department of Philosophy, Universität Konstanz, Konstanz, Germany

Let me first state that I like Antti Revonsuo's discussion of the various methodological and interpretational problems in neuroscience, with its demonstration of the ways in which careful and methodologically reflective scientists must proceed in this fascinating field of research. I have nothing to add to that discussion. Furthermore, I am very sympathetic toward Revonsuo's call for a philosophy of neuroscience that stresses foundational issues but also focuses on methodological and explanatory strategies.[1] Admittedly, I am myself a trained physicist and not a neuroscientist; I will therefore use examples from physics to illustrate my points.

My comments address some of the main philosophical theses of Revonsuo's chapter (especially the ones I disagree with) and are divided into three sections. The first deals with the method of visualization, which I think is not an acceptable explanation. The second section discusses different views of explanation in the light of neuroscience. Here I focus especially on the presuppositions of the two dominant theories of explanation and stress, contra Revonsuo, the idea of coherence. Finally I address the issue of conflicting explanatory strategies for the same phenomenon in cognitive neuroscience and the suggested methodological consequences Revonsuo draws from this.

Explanation and Visualization

In several parts of his chapter, Revonsuo describes and praises the use of visualizations in neuroscience. Two types of visualizations deserve special attention: (1) the exposition of neural mechanisms and (2) brain imaging and mapping methods. There is no doubt that these visualizations are important tools in the actual research process: they help scientists grasp complicated systems, they are heuristically useful, they represent data, and they serve various didactic purposes (Ruse 1991; Wimsatt 1991). Many scientists have shared the experience of "knowing immediately what is going on" once they see a well-constructed diagram. Evidently this is why biology books are full of them.

I disagree, however, that visualizations provide or facilitate explanations. Revonsuo argues for the explanatory power of visualizations when he claims—citing Bechtel and Richardson (1992) approvingly—that idealized models "may be only partially (if at all) clothed in linguistic representations; instead, all kinds of diagrams and figures can often best depict the component structures of, and their mutual interactions within, the biological system in question." Diagrams and figures are therefore models, and models are taken to be explanatory. I would object that, in order to make sense of a depiction of a mechanism, one needs all sorts of theories and theoretical models in the background. Diagrams and figures refer to, or at least hint at, the theoretical treatment in the background by means of various conventions shared by the users of the depictions (Bailer-Jones 2001). Without theories and models and the pictorial conventions pointing to them, it is not at all clear what a diagram *means* and how it relates to the phenomenon to be explained. Besides, taking visualizations to be explanatory cannot account for the common intuition that explanations can be true or false. It is precisely because visualizations employ some quite arbitrary, though convenient, conventions that they elude the categories of truth and falsity. Visualizations can only be more or less useful for a certain purpose. In the case of brain images, Revonsuo himself points out how much theoretical knowledge is required to interpret correctly the pictures obtained. Moreover, measurement methods such as positron emission tomography (PET) and functional magnetic resonance imaging (fRMI) often presuppose certain key assumptions (e.g., modularity) that may not hold in nature. Taking these images literally as explanations would therefore be quite a dubious procedure. Let me therefore examine what scientific explanations really are.

Mechanisms and Coherence

Most philosophers agree that a major aim of science is to explain phenomena. Although the concept of explanation is somewhat vague, an explanation should show (1) how the phenomenon under consideration reached its present state and (2) how it fits into a larger theoretical framework. An acceptable answer to the first question produces *local understanding;* dealing successfully with the second question provides us with *global understanding.* Although these two requirements are not mutually exclusive, it remains to be seen if both can be fulfilled by the same scientific theory or model. Philosophical theories of explanation therefore usually concentrate on one of these requirements—a task that turns out to be hard enough, as the controversy that has raged over the past four decades or so impressively shows (Salmon 1989).

Although it is generally agreed that pragmatic considerations do play an enormously important role in scientific explanations, Revonsuo considers only the causal-mechanical account and the unification account. As I show in the next section, this neglect of pragmatic considerations is somewhat unfortunate. According to the causal-mechanical account, pioneered by Salmon (1998) and others, a phenomenon is explained by uncovering a "hidden [mechanism] by which nature works." There are several variants of this account to be found in the literature, such as the proposals by Bechtel and Richardson, Humphreys, Salmon, and Woodward.[2] Revonsuo, however, bases his reflections only on the specific account put forward by Bechtel and Richardson (1993) in a series of publications.

According to the unification account, developed by Friedman and elaborated by Kitcher (1989), a successful explanation fits the explanandum in a coherent way within a general framework. This view, which is a distant descendant of Hempel and Oppenheim's famous original account, supports the intuition that something is explained if it is covered by general principles. But general principles and universal laws are rare in neuroscience, which is why Revonsuo hastily concludes that unification does not play a role in this field of research. It should be noted, however, that universal principles frequently do play a role in the ordinary explanatory business of the neurosciences. This is demonstrated by the observation that one whole chapter (of seven individual contributions) in the authoritative anthology *The Cognitive Neurosciences* (Gazzaniga 1995) is devoted to "Evolutionary Perspectives." In the introduction to this chapter, section editors John Tooby and Leda Cosmides point out that "evolutionary biology has a great deal to offer cognitive neuroscience. Because human and nonhuman brains are evolved systems, they are organized according to an underlying evolutionary logic. By knowing what adaptive problems a species faced during its evolutionary history, researchers can gain insight into the functional circuitry of its neural architecture" (Tooby and Cosmides 1995, 1181). Besides, neuroscientific explanations ought to be consistent with all fundamental principles of physics (e.g., conservation laws). Though these principles may not be of direct help in finding causal mechanisms, serious problems arise if a proposed mechanism violates them. Fundamental laws and principles set restrictions that may eventually even suggest a detailed "local" explanation. Another aspect of the unification account is even more important. Unlike the causal-mechanical account, the unification account stresses the role of coherence considerations in science. I return to this issue subsequently.

Revonsuo reminds us that both approaches to scientific explanation make assumptions about the structure of the world that may not hold. David Lewis once asked what happens if nature is not unified. This, at first sight, seems to be a difficulty for the unificationist, and Kitcher (1989) addresses this problem in detail. Revonsuo now challenges the causal-mechanical account by asking what would happen if the assumptions of decomposition and localization did not hold. *Decomposition* here means that the system is composed of modular subsystems, *localization* that the subfunctions within these parts can be identified (Bechtel and Richardson 1993). Indeed the cases Revonsuo presents suggest that decomposition and localization may not be feasible in the brain.[3] This is not to suggest that the complete causal-mechanical program is at stake, only that the variant of this program Revonsuo adopted might be too narrow. Presumably a more "liberal" account of the causal-mechanical program could fully avoid these problems.

First, I do not see why it is essential to the causal-mechanical program that subfunctions be localized in well-defined and spatially separated regions. In physics, parts that form functional units are often spatially separated over large distances. A typical example is super-conductivity. The two correlated electrons that form the so-called Cooper pairs (i.e., the effective degrees of freedom of a superconductor) may be localized at opposite ends of the superconductor. A "quasi-mechanical" explanation of superconductivity on the basis of the properties of Cooper pairs can nevertheless be achieved.[4] Second, Revonsuo argues that a system is decomposable if the "causal interactions *within* the subsystems [are] more important than those *between* the subsystems." This does not seem to apply in any field of physics either. According to quantum chromodynamics, the fundamental theory of strong interactions, quarks are elementary and do not have a substructure. They do, however, heavily interact inside hadrons (e.g., protons, neutrons, pions) at low and intermediate energies. In fact it is not possible to decompose the system in the laboratory and to localize individual quarks (Hartmann 1999). It is, however, possible to write down equations for the detailed mechanism that facilitates the strongly attractive interaction between the quarks.

In sum, I do not see that the causal-mechanical program is in trouble in neuroscience. The examples Revonsuo discusses merely suggest that the program must be better adapted to the empirical facts. Since our theories of explanation (and especially the causal-mechanical account) make strong assumptions about the world, it is no surprise that we run the risk that some of these assumptions will turn out not to hold as the scientific endeavor progresses. As far as I can see, a wider account (such as the one presented by Machamer et al. 2000; see also Craver and Darden's contribution to this volume) seems to be able to deal with the problematic cases Revonsuo presents. According to Machamer et al. (2000, 3), mechanisms are "entities and activities organized such that they are productive of regular changes from start or setup conditions to finish or termination conditions." This is not the place to flesh out this characterization in detail. It is only important to note that a mechanism in this framework simply tells us how a system evolved in time, and this seems to apply to Revonsuo's critical cases.

Having defended the causal-mechanical program against the charges brought by Revonsuo, I now present some of my own critical arguments. I consider all variants of the causal-mechanical program I know of to be incomplete because they do not sufficiently stress the important role of *coherence* considerations in the process of establishing specific mechanisms. A proposed mechanism must cohere with our accepted background knowledge. Here it is important to note that different research programs in a preparadigmatic phase of a science (such as cognitive neuroscience) may incorporate different beliefs into their respective background knowledge. Although some beliefs may be taken to be noncontroversial by all competing programs, some may be accepted by one program and dismissed by others (such as views about the explanatory importance of the neural level or the epistemological status of folk psychology). But once a chunk of background beliefs is provisionally accepted, new beliefs in this framework should cohere with it.

Coherence is a term that is notoriously hard to define. Some plainly identify it with logical consistency. But logical consistency seems neither necessary nor sufficient for (approximate) coherence since there are usually great uncertainties in our background beliefs (especially in neuroscience, as Revonsuo stresses in his chapter). So if some propositions of a belief system are uncertain, a contradiction resulting from integrating a new proposition into the system would not destroy the coherence of the whole system. Besides, logical consistency does not seem to be a very good guide in pointing to a desired mechanism. Too many mechanisms do the job of bringing the system from here to there, but not all of them are accepted—for good reasons.[5]

Although a new mechanism can suggest a radically new aspect of our world, it will still be linked to other parts of our belief system. The question of how well this new mechanism coheres with the rest of this system is therefore an important one. In order to make this claim precise, it is necessary to plunge into the deep epistemological waters of defining what coherence means. Here is what Lawrence BonJour (1985, 93) has to say:

> What then is coherence? Intuitively, coherence is a matter of how well a body of belief "hangs together": how well its component beliefs fit together, agree or dovetail with each other, so as to produce an organized, tightly structured system of beliefs, rather than either a helter-skelter collection or a set of conflicting subsystems. It is reasonably clear that this "hanging together" depends on the various sorts of inferential, evidential, and explanatory relations which obtain among the various members of a system of belief, and especially on the more holistic and systematic of these.

Following this line of thought, we would accept a proposed mechanism if it makes a given system of beliefs more coherent, or at least if it does not make it less coherent. This seems to be intuitively clear, and

Revonsuo himself discusses a couple of examples in which coherence considerations play a role. It is obvious, for example, that the models of neural systems at several levels of description must cohere. Revonsuo here mentions "synapses and synaptic transmission, single-neuron morphology and electrophysiology, neural connectivity and organization in sensory systems and the central nervous system, cytoarchitectonics of the cerebral cortex, macroanatomy of the brain, and so forth." It would indeed be a miracle if models of so many interrelated levels fit together without using coherence as an important constraint in theory construction. Another example of the role of coherence considerations in the neurosciences is the above-mentioned interpretation of pictures obtained by PET or fMRI. Acceptable results of such measurements should cohere with the other assumptions of the "experimental paradigm." Since "any one of [these assumptions] might be wrong," this turns out to be a difficult task.

Although all this seems to be intuitively clear, and scientists use such principles in their daily work, it is nevertheless a problem in formal philosophy to assign a quantitative measure (say, a number between zero and one) to the coherence of a belief system. I suggest that this is best done within a probabilistic framework. Luc Bovens and I have shown elsewhere that such a qualitative measure of coherence can be obtained if one specifies coherence to be a confidence-boosting property of a set of beliefs and uses the mathematical theory of Bayesian networks, which is well known in artificial intelligence research (Bovens and Hartmann forthcoming).

I propose that the search for coherence is an important guiding principle in finding acceptable explanations in neuroscience.[6] Most of these explanations might indeed be causal-mechanical (although I doubt that all are), but if a suggested mechanism does not fit within a set of background assumptions it will have a hard time being accepted by the scientific community. Perhaps there is nothing more to gain from science than a picture of the world that is as coherent as possible. Even if the causal-mechanical account really must be abandoned at some point, the idea of coherence will still play a dominant role in scientific theorizing.

I therefore doubt that there are really two alternative views of explanation that may complement each other and even coexist in science. Recall Salmon's (1998, 73–74) story of the friendly physicist who asks a young boy sitting near him in an airplane awaiting takeoff what the helium-filled balloon the boy is holding will do when the airplane accelerates for takeoff. Perhaps somewhat counterintuitively, the balloon moves forward. Salmon states that there are two equally acceptable explanations for the phenomenon. The causal-mechanical explanation relies on an intuitive story about the movement of the gas molecules, while the explanation-as-unification applies Einstein's equivalence principle. Again I stress that local laws, such as the ones that govern the behavior of gas particles, would not be accepted as explanatory if they contradicted general principles such as the equivalence principle and if they were not part of a coherent larger framework. It is the interaction between these two approaches (a bottom-up approach and a top-down approach) that seems best to characterize science and its ability to explain. General considerations frequently help us to obtain "local" knowledge, and the analysis of specific mechanisms may provide "global" knowledge.

Co-Evolution and the Place of Psychology

According to Revonsuo, there is a deep confusion at the foundation of the new discipline of cognitive neuroscience because conflicting explanatory strategies crash at the interface of the old disciplines of cognitive science and neuroscience. While neuroscience aims to explain mental or cognitive levels in a mechanistic way, cognitive science is committed to functionalism and the autonomy of psychology. In Revonsuo's view functionalism implies "the total independence of psychological and neural levels of description." This is certainly too strong a claim, since even if there are multiple ways to realize an algorithm, it must still be shown or made plausible that the human brain is able to do so.[7]

Even if we accept, for the sake of argument, that there is a conflict (after all, at the end of the day eliminative-materialists want to get rid of folk psychology!), we are still left with the question of what to do in this situation. Revonsuo has a radical proposal: he suggests that we abandon the cognitivist-functionalist strategy and argues for a "full-scale" mechanistic program to be applied to cognitive neuroscience.[8]

Again I consider this to be too radical a proposal, and I am not sure if Revonsuo is serious about what he writes. First, Revonsuo himself pointed out that the causal-mechanical program might be in trouble. Second, it is not clear to me that the evidence for the causal-mechanical account is so overwhelming, and the evidence for the cognitive science account so poor, that one must abandon the latter and follow the former. Third, and most importantly, research programs that evolve in parallel can be of great advantage to each other. Churchland (1986, 384) explains why: "Theories at distinct levels often co-evolve . . . , as each informs or corrects the other, and if a theory at one stage of its history cannot reduce a likely candidate at a higher level, it may grow and mature so that eventually it does succeed in the reductive goal. In the meantime the discoveries and problems of each theory may suggest modifications, developments, and experiments for the other, and thus the two evolve towards a reductive consummation." This quote shows that even a strong proponent of an eliminativist-materialistic ontology can live with methodological pluralism (and indeed advise it). In any case, Revonsuo's thesis seems to me to be by far too strong.

One may wonder if the pluralism I just defended is a problem for the aim formulated in the last section, the attainment of coherent theories. Obviously, if we include cognitive science and neuroscience in our theory of the world, the resulting system will not be considerably coherent. Having a measure of coherence might suggest, pace Revonsuo, only following the program that seems to be the more coherent of the two. In accordance with the idea of coevolution, I would, however, not advise this. We must live with the fact that some sciences are still in a preparadigmatic stage, and in this situation it is best to let both programs grow, allowing both to profit from cross-fertilization. Coherence now should play a role only within a given framework (say, cognitive science, plus background knowledge from other sciences, plus experimental data, and so forth). The competing research programs should not be taken into account here. Inconsistencies between different research programs do not matter at this stage of theory development. This, in a way, is a pragmatic component. Once we have chosen a certain framework, we can reach for coherent theories. The question of when a research program must be abandoned of course requires further investigation.

What, finally, is the exact place of psychology? This will remain an open question until we have a satisfactory account of our cognitive functions. Until then, there is nothing wrong with a plurality of different explanations for the same phenomenon.[9] It is legitimate to approach a given subject from different directions without worrying too much about their mutual consistency. In this same spirit Patricia

Churchland (1986) once advised us to follow the research principle "Let a thousand flowers bloom."

NOTES

I thank Daniela Bailer-Jones for very helpful editorial and substantive suggestions.

1. It should be noted here that Patricia Churchland does, in contrast to what Revonsuo implies, indeed discuss methodological issues of neuroscience in her *Neurophilosophy* (1986). In this book (and also in subsequent publications; see Churchland and Sejnowski 1992) she defends—following the work of Wimsatt (1976)—a pluralistic methodology dubbed coevolution of theories (see pp. 284–86 and 362–76). I will return to this theme in the section entitled "Co-Evolution and the Place of Psychology." It is therefore not correct to claim that "neurophilosophy is regarded merely as an expression of an eliminativist-reductionist program in the philosophy of mind" because the methodological strategy of coevolution, which Churchland defends, leaves considerable space for different programs to develop.

2. The views of the last three authors are presented in Kitcher and Salmon (1989).

3. This makes Revonsuo's suggestion that only the causal-mechanical program should be considered in the study of consciousness somewhat surprising.

4. For an interesting discussion of indeterministic mechanisms see Ackermann (1968, 1969).

5. Bailer-Jones (2000) develops the role of causal mechanisms in a similar direction, but misleadingly talks about consistency rather than coherence.

6. The relation between the notions of coherence and explanation is also discussed in Bartelborth (1999).

7. Another argument against the multiple-realizations argument is given in Bechtel and Mundale (1999).

8. Revonsuo advises this strategy even for the case of consciousness, although he formulates only vaguely his belief that the causal-mechanical account is superior.

9. For a defense of this view for biology, see Mayr (1997).

REFERENCES

Ackermann, R. 1968. "Mechanism and the Philosophy of Biology." *Southern Journal of Philosophy* 6: 143–51.
——. 1969. "Mechanism, Methodology, and Biological Theory." *Synthese* 20: 219–29.
Bailer-Jones, D. M. 2000. "Modelling Extended Extragalactic Radio Sources." *Studies in History and Philosophy of Modern Physics* 31B: 49–74.

———. 2001. "Sketches as Mental Reifications of Theoretical Scientific Treatment." In M. Anderson, B. Meyer, and P. Olivier, eds., *Diagrammatic Representation and Reasoning*. London: Springer-Verlag.

Bartelborth, T. 1999. "Explanatory Coherence." *Erkenntnis* 50: 209–24.

Bechtel, W., and J. Mundale. 1999. "Multiple Realizability Revisited: Linking Cognitive and Neural States." *Philosophy of Science* 66: 175–207.

Bechtel, W., and R. C. Richardson. 1992. "Emergent Phenomena and Complex Systems." In A. Beckermann, H. Flohr, and J. Kim, eds., *Emergence or Reduction? Essays on the Prospects of Nonreductive Physicalism*. Berlin: de Gruyter, 257–88.

———. 1993. *Discovering Complexity: Decomposition and Localization as Scientific Research Strategies*. Princeton, N.J.: Princeton University Press.

BonJour, L. 1985. *The Structure of Empirical Knowledge*. Cambridge, Mass.: Harvard University Press.

Bovens, L., and S. Hartmann. Forthcoming. "The Riddle of Coherence." Preprint, Boulder and Konstanz.

Churchland, P. S. 1986. *Neurophilosophy*. Cambridge, Mass.: MIT Press.

Churchland, P. S., and T. Sejnowski. 1992. *The Computational Brain*. Cambridge, Mass.: MIT Press.

Gazzaniga, M., ed. 1995. *The Cognitive Neurosciences*. Cambridge, Mass.: MIT Press.

Hartmann, S. 1999. "Models and Stories in Hadron Physics." In M. Morrison and M. Morgan, eds., *Models as Mediators: Perspectives on Natural and Social Science*. Cambridge: Cambridge University Press, 326–46.

Kitcher, P. 1989. "Explanatory Unification and the Causal Structure of the World." In Kitcher and Salmon 1989, 410–505.

Kitcher, P., and W. Salmon, eds. 1989. *Scientific Explanation*. Minneapolis: University of Minnesota Press.

Machamer, P., L. Darden, and C. F. Craver. 2000. "Thinking about Mechanisms." *Philosophy of Science* 67: 1–25.

Mayr, E. 1997. *This Is Biology: The Science of the Living World*. Cambridge, Mass.: Harvard University Press.

Ruse, M. 1991. "Are Pictures Really Necessary? The Case of Sewall Wright's 'Adaptive Landscapes.'" In A. Fine, M. Forbes, and L. Wessels, eds., *PSA 1990*, vol. 2. East Lansing, Mich.: Philosophy of Science Association, 63–77.

Salmon, W. 1989. *Four Decades of Scientific Explanation*. Minneapolis: University of Minnesota Press. (Also contained in Kitcher and Salmon 1989, 3–219.)

———. 1998. *Causality and Explanation*. Oxford: Oxford University Press.

Tooby, J., and L. Cosmides. 1995. "Evolutionary Perspectives." In Gazzaniga 1995, 1081–286.

Wimsatt, W. 1976. "Reduction, Levels of Organization, and the Mind-Body Problem." In G. Globus, G. Maxwell, and I. Savodnik, eds., *Consciousness and the Brain*. New York: Plenum Press, 199–267.

———. 1991. "Taming the Dimensions: Visualizations in Science." In A. Fine, M. Forbes, and L. Wessels, eds., *PSA 1990*, vol. 2. East Lansing, Mich.: Philosophy of Science Association, 111–35.

5

Cognitive Neuroscience

Relating Neural Mechanisms and Cognition

William Bechtel
Philosophy-Neuroscience-Psychology Program, Department of Philosophy,
Washington University, St. Louis, Missouri

In the 1970s and 1980s, the study of cognition and the study of the brain were carried out largely in isolation from each other. Each was the subject of its own interdisciplinary cluster, which had only recently taken shape—cognitive science focusing on cognition, neuroscience on the brain. But in the 1980s the seeds were already being sown for something far grander—an integration of cognitive science and neuroscience in cognitive neuroscience.[1] In the 1990s cognitive neuroscience matured rapidly, and at the start of the new millennium it is positioned as one of the most vital fields of inquiry. Part of my task is to analyze how cognitive neuroscience has reached this status and how it proposes to connect neuroscience and cognition.

But there is a second part to my task. Cognitive neuroscientists, to a large degree, proceed as if there had never been a mind-body problem. To a Cartesian-minded philosopher—or even to a 1980s functionalist in the philosophy of mind—this must seem deeply perplexing. The cognitive properties of the mind have seemed to such philosophers to be radically disconnected from those of the brain. How is it that those enmeshed in the study of cognitive processes in the brain have not been ensnared by the mind-body problem? In large part, as I try to show, it is because the explanatory framework they are adopting is one that naturally relates cognitive and neural processes. I attempt to illustrate this linkage through a brief case study of the history of research on visual processing. (It is worth noting that the history of research on vision is much longer than the modern enterprise of cognitive neuroscience. This hundred-year-long research endeavor, which by now has exhibited a substantial degree of success, is not a product of cognitive neuroscience, but it provides an exemplar for the more recent endeavors in cognitive neuroscience.)

The Emergence of a New Interdisciplinary Research Cluster

As interdisciplinary research clusters, both cognitive science and neuroscience trace their institutionalization to the period 1950–70. Both involved the drawing together of a wide variety of separate disciplines. The term *neuroscience* was introduced as part of an attempt to relate a variety of scientific investigations of the brain, ranging from the molecular and cellular levels to the systems and behavioral levels, which had previously been carried on in separate disciplines. In 1962 Francis Schmitt developed a proposal and received approval and funding for a Neurosciences Research Program at MIT, which became a major catalyst for the development of a new interdisciplinary cluster in the neurosciences. In 1970 the Society for Neuroscience was created; starting with 500 members, it has grown to over 25,000 within thirty years.

Cognitive science—an interdisciplinary cluster principally involving parts of psychology, linguistics, and computer science and to a lesser degree elements of philosophy, anthropology, and neuroscience—arose in part out of the repudiation across a variety of disciplines of the strictures of behaviorism. In linguistics this took the form of the development of generative grammars à la Chomsky; in psychology it involved developing information processing accounts of internal mental processes such as encoding, storage, and retrieval operations in memory and analyses of procedures employed in problem solving; and in computer science it took the form of computational models of reasoning and problem solving. Although pioneering efforts in all of these areas occurred in the 1950s, they reached maturity in the 1960s and 1970s. In a pattern that was to be repeated in cognitive neuroscience, the rapidity of development in cognitive science was facilitated by funding from private foundations, especially the Sloan Foundation, which maintained an initiative in cognitive science for ten years beginning in the mid-1970s. During this time Sloan provided funding to establish centers of cognitive science at a number of U.S. institutions, including MIT, Stanford, the University of California at Berkeley, the

University of California at San Diego (UCSD), and the University of Pennsylvania. Supported in part by the Sloan grant to UCSD, the Cognitive Science Society was born at a conference in August 1979 (Bechtel et al. 1998).

In principle both the Society for Neuroscience and the Cognitive Science Society could each have incorporated the domain of the other. But in fact each tended to exclude the other. On the neuroscience side, this was a consequence of rapid success in cellular and molecular areas, especially with the development of techniques for intervention at the genetic level, which largely overshadowed work at the behavioral level. For cognitive science, this was partly a consequence of focusing on computational models, especially ones employing operations performed on symbolic or language-like strings. Although such models, whether developed in psychology or artificial intelligence, could be implemented on von Neumann computers, it was less obvious how they could be implemented in neural tissue.

Starting in the 1980s, the gulf between the cognitive sciences and neurosciences began to diminish. On the one hand, neuroscience researchers trying to determine precisely what information was carried by specific neural activity (which could be registered through such techniques as single-cell recording) recognized the need for much more sophisticated behavioral protocols—precisely the sort of thing that cognitive psychologists had become adept at developing. Cognitive psychologists had been forced by the very limited range of dependent variables available to them (principally reaction times and error patterns) to be clever in the design of behavioral tasks to maximize the chance of procuring informative data. But at the same time, some cognitive psychologists recognized the limitations of such dependent variables and eagerly sought new tools that might provide more direct information about the cognitive operations the brain was actually performing (Posner and McLeod 1982). In 1986 Joseph E. LeDoux and William Hirst produced a book, *Mind and Brain: Dialogues in Cognitive Neuroscience,* that drew together neuroscientists and psychologists to review the state of the art in their own discipline and to respond to the corresponding reviews of the other on four topics: perception, attention, memory, and emotion. These reviews reveal not only the emergence of a desire on the part of some practitioners in each area to be able to draw upon the resources of the other but also the large differences between the investigations pursued. To highlight just one example of the differences: while most cognitive psychologists conducted their studies on adult humans, most neuroscientific research was conducted on other species, where ethical constraints did not prevent the insertion of electrodes into brains or the making of experimental lesions.

Besides the growing interest of both cognitive scientists and neuroscientists in each other's field, the development of a new technology that could image activity in human brains—positron emission tomography (PET)—served as a major impetus spanning the fields. PET was primarily developed to study the hemodynamics of the brain, a topic of particular concern to neurologists concerned with vascular diseases of the brain. Early in its development, though, certain researchers, such as Marcus Raichle, recognized its potential for studying brain areas involved in mental activities.[2] In a first demonstration, Raichle and his colleagues showed that simple sensory and motor tasks would cause increased blood flow in primary sensory and motor areas (Fox et al. 1986). But it was a long way from this beginning to developing a tool that could reveal brain activity involved in higher cognitive functions such as recognizing an object, processing the semantics of language, or encoding information into memory. A first step in that direction stemmed from an interesting quirk in funding. The James S. McDonnell Foundation made a $40 million grant to the Washington University School of Medicine to create a center for the study of higher brain function. To ensure a focus on higher brain function, a condition of this grant was that a psychologist be employed in the research. This stipulation resulted in Michael Posner, a leader in the use of reaction times to identify mental processes (Posner 1978), being brought to Washington University.

A common (but criticized; see Sternberg 1969) approach to reaction time studies was to subtract the time required to perform one task from that required to perform another that was thought to differ from the first in the inclusion of one additional cognitive operation, so as to determine the time required for performing the additional operation. Raichle and Posner now adapted this approach to PET, subtracting the activations generated by performing one cognitive task from those generated by another that was thought to entail one additional mental operation. In a now classic experiment, they subtracted the activation produced in merely reading a noun aloud from the activation produced by generating and pronouncing a verb suggested by a noun

(Petersen et al. 1989). As PET developed, and with the subsequent introduction of functional magnetic resonance imaging (which employs an endogenous magnetic signal of rate of blood flow and offers better spatial and temporal resolution), neural imaging[3] has become an attractive research method for both psychologists and neuroscientists, drawing the two research areas closer together.

The now widely adopted term for the interdisciplinary area, *cognitive neuroscience,* seems to have been coined by neuroscientist Michael Gazzaniga in dialogue with psychologist George Miller.[4] Gazzaniga has played a variety of important roles in the further development of cognitive neuroscience. In 1986 he submitted a proposal for research in cognitive neuroscience to the McDonnell Foundation, which had just the previous month decided "to develop some specific programs to support research linking the biological with the behavioral sciences." In March 1987, the foundation convened a meeting of eight senior scientists (in addition to Gazzaniga and Miller, they included Posner and Raichle, Emilio Bizzi, Steven Hillyard, Terry Sejnowski, and Gordon Shepherd) "to discuss the promises and pitfalls of funding research in the field of cognitive neuroscience and to begin the process of developing program criteria and requests for proposals." Concluding that there were "too many open questions, theoretical tangles and potential misunderstandings separating the two critical specialties—neuroscience and psychology—to proceed immediately," this panel recommended the creation of five study panels composed of investigators from neuroscience, linguistics, artificial intelligence, and philosophy to examine the state of research on emotion and cognition, memory and learning, attention and perception, higher cognitive processes, and motor control. These panels met over the succeeding two years. In addition to holding sessions in which they brought in leading investigators to arrive at a sense of the state of the field, the panels supported preliminary projects in each of their areas.[5] In addition, in 1988 the foundation began funding a summer institute in cognitive neuroscience (the first institute was co-directed by Stephen Kosslyn together with Posner and Shepherd; subsequent annual institutes have been directed by Gazzaniga).

On the basis of the study group reports, the McDonnell Foundation began a collaboration with the Pew Charitable Trusts that resulted in a major investment in cognitive neuroscience. One of their strategies was to create eight McDonnell-Pew centers for cognitive neuroscience at universities at which the core of a cognitive neuroscience community was already in place: the University of Arizona, Dartmouth College and Medical School,[6] the Johns Hopkins University, the Montreal Neurological Institute, MIT, Oxford University, the University of Oregon, and UCSD (in conjunction with the Salk Institute and the Scripps Clinic). Johns Hopkins was dropped in 1993 (partly as a result of key personnel moving elsewhere), but the other centers continued to receive substantial McDonnell-Pew support through 1998.

Largely as a result of differences in the existing structure at each institution, each of the centers developed a somewhat different perspective on cognitive neuroscience. For example, MIT had had a Department of Psychology and Brain Science since 1962 (renamed the Department of Brain and Cognitive Sciences in 1986) but had not really integrated the two elements of the department. The McDonnell-Pew support allowed for the development of an integrated track in which graduate students took core courses in both neuroscience and cognitive science, had advisors from at least two components of the department (systems neuroscience, cognitive science, and computational modeling), and pursued dissertation projects linking these areas. Building from a foundation in middle-level vision and middle-level motor control, the department used the opportunity to expand into language processing and higher-level vision. Both UCSD and Oxford had large contingents in neuroscience and cognitive science prior to receiving the McDonnell-Pew support, but the two groups had been largely isolated in different departments or schools. At Oxford the support permitted expansion of the traditional three years of support for the Ph.D. to four, and training that involved both psychology and neuroscience. The UCSD program emphasized integrative activities, such as retreats and regular interactions among McDonnell-Pew fellows (graduate students and postdoctoral fellows in various departments who received McDonnell-Pew support), thereby training researchers who had a much expanded perspective on their research problems and who were able to integrate easily the traditional tools of several disciplines.

The McDonnell-Pew support began the process of institutionalizing cognitive neuroscience at a number of universities and building a community of cognitive neuroscientists.[7] The community building also involved the founding of a new journal (the *Journal of Cognitive Neuroscience*) in 1989 and of the Cognitive Neuroscience Society in

1994 (Gazzaniga played the leading role in both of these initiatives). In recent years, national funding agencies, such as the National Science Foundation and the National Institutes of Health in the United States and the Medical Research Council in England, have begun to provide major support for research in cognitive neuroscience.

Judged by such criteria as the inflow of researchers, the creation of new journals, and the procurement of financial resources, cognitive neuroscience is clearly a thriving endeavor. But what is distinctive about the project of cognitive neuroscience? Is something emerging from the collaboration of fields that would not have emerged from one or the other of the contributing fields on its own? One quite plausible view is that the marriage of cognition and neuroscience merely provides a means to test the accuracy of cognitive models by determining whether the individual cognitive processes specified in cognitive models are actually performed in the brain. Given the under-constrained nature of cognitive investigations, this would certainly increase the evidential basis of cognitive models. While this is one benefit of the marriage of cognition and neuroscience, there is another, potentially more valuable, contribution—the integration of cognitive and neuroscientific research tools may provide insight into the basic cognitive processes themselves. In contrast to the picture painted by eliminativists who propose that cognitive theories will be supplanted by new accounts developed on the basis of neuroscience theories, the result is not one of neuroscientists supplanting cognitive scientists but one of true integration, in which the resulting models relate neural processes and cognition.

As I noted earlier, the prevailing metaphysical backdrop against which cognitive neuroscience has emerged has been a form of functionalism that emphasizes the autonomy of cognitive and neural inquires. According to functionalism, a belief (the sort of cognitive state that has figured prominently in philosophical writings about cognition, but seldom in psychology or the other cognitive sciences) is identified in terms of its relation to other mental states (e.g., other beliefs, desires). Mental states are thus construed as constituting a network, with the position of each state in the network determining its identity. Functionalists generally have presented themselves as rejecting the identity theory, according to which a given mental state is identified with a particular brain state.[8] So far a distinctively cognitive neuroscience stance on the mind-body problem has not been presented.

However, such a stance, which endorses aspects of both identity theory and functionalism, is implicit in cognitive neuroscience.

Such a perspective is already in place in other domains of biology, where an emphasis on function (such as that pursued in physiology) is also often contrasted with an emphasis on structure (such as that pursued in anatomy). In these cases, discussion of the contrast is combined with the recognition of an intimate relationship between the structural and functional perspectives. Why has such a relationship not been countenanced in the case of cognition and the brain? Perhaps a major factor is the attractiveness of the metaphor of the distinction between hardware and software, in which cognitive inquiries are directed at identifying the software and neuroscience inquiries are directed at specifying the hardware. A popular argument for such autonomy has been the alleged multiple realizability of mental states, according to which it is first maintained that the same cognitive function or program can be realized in different neural structures or hardware. Hunger, to use Putnam's (1967) example, might be realized by different neural processes in octopuses and human beings. The conclusion is then advanced that the neural structure or hardware is uninformative as to the cognitive operations or program (Fodor 1974). This invocation of the language of hardware and software, however, has misled philosophers as to what neuroscience has been about—engendering the view that its focus has been limited to physical structures.[9]

In fact, neuroscience research often addresses function, including mental or cognitive function. The mapping of brain areas, from Brodmann and other turn-of-the-century researchers through to contemporary researchers such as van Essen and Feldman, has been directed toward and guided by functional considerations at a variety of levels.[10] The principal limitation has been that, until recently, the main tool for neural investigation of cognitive function in humans has been the analysis of deficits resulting from naturally occurring lesions, which often do not conform to either anatomical or functional boundaries. In addition, the deficits are often rather gross, not helping to illuminate specific cognitive operations. The techniques that have proven most useful in isolating more micro-level functions—surgically induced lesions and single-cell recording—have generally been restricted to use with nonhuman animals. The obstacles to functional analysis in neuroscience are practical, not principled, and at lower levels (e.g., at the level of nerve transmission) functional analyses are just as widespread as concern for structure.

Recognizing the role of functional considerations in neuroscience already helps to close the apparent ontological gap between brain processes and mental processes (and also, although I cannot develop this topic here, provides the basis for answering the multiple-realizability objection; see Bechtel and Mundale 1999). But we can close the gap even further by identifying a common explanatory framework employed in both the cognitive sciences and the neurosciences— the framework of mechanistic explanation (Bechtel and Richardson 1993; Craver and Darden this volume; Machamer et al. 2000). A necessary step in developing a mechanistic explanation is identical to what the functionalists in philosophy have proposed (Dennett 1971): the decomposition of cognitive performance so as to identify the component operations within the mind.[11] An explanation then consists in a proposal of how, through the joint and coordinated performance of these component operations, the behavior one wants to explain is generated.

The goal of research is not to provide only a possible account of how a behavior arose, but the account that correctly characterizes the cognitive agent. This requires providing evidence that the component tasks invoked in the explanation are among those actually being performed in the system and thus that one has the right decomposition. Richardson and I have used the term *localization* for the securing of evidence that something performs the task. Often the agent performing the task is spatially localized, but sometimes it is distributed over several components of the system. For the most part, cognitive scientists have been restricted to behavioral measures (reaction times and error patterns) for their evidence, but their ready willingness to embrace neuroimaging shows that this restriction was a question of practice, not principle.

When function is viewed in the context of mechanistic explanation, it is quite compatible with linked cognitive and neural perspectives. Moreover, the structures that are appealed to as the performers of the component tasks are themselves doing things that need to be explained. At a lower level, therefore, the structure is itself construed functionally, and one turns to yet lower-level structures to explain it. The functionalism actually operative in cognitive neuroscience is thus one that is compatible with an identity theory (Bechtel and McCauley

1999). But the mechanistic approach is not reductive in the classical philosophical sense, since it is not built around the notion of laws or the idea that laws describing interactions at one level are derivable from those at another.[12]

In the foregoing construal, I have characterized a top-down approach in which one tries to envision how a system might be organized so as to carry out a particular task and then seeks evidence supporting that proposal. But one can also proceed in a more bottom-up fashion by first identifying components of a system, determining how they operate, and building from this information up to an overall account of the mechanism that performs the function. One advantage of this approach is that one may discover a decomposition that one had not anticipated. This approach is exemplified in the history of research that has unraveled basic features of the design of the mechanism involved in primate vision, which I briefly describe in the next section. One thing to emphasize, though, is that while it has been neuroscientific studies of actual brain components that have played the major role in discovering the organization of this mechanism, the neuroscientific investigations have themselves been focused on accounting for the higher-level activity of seeing, and much of the evidence used to develop these models is behavioral.

Prototypical Cognitive Neuroscience Research: The Development of a Mechanistic Model of Visual Processing

In this section I present a brief history of the major steps in the development of the contemporary account of the neural mechanisms of vision. This research offers a Kuhnian exemplar in the field of cognitive neuroscience—a success story to be emulated. It also exhibits how neural and behavioral perspectives had to be integrated in order to discover the mechanisms of vision.

Identifying an Initial Locus of Control

Even to begin an investigation of a process such as seeing, one has to start with naively simple assumptions. Thus, in developing explanations of how a complex system performs a function, researchers often attempt first to localize *that* function in a particular part of the system (Bechtel and Richardson 1993). Such attempts often become controversial when there are competing candidate localizations, and research is devoted to adjudicating between them. In the late nineteenth century such a conflict arose between a proposal to localize visual processing in the occipital cortex—first advocated by Bartolomeo Panizza (1855) on the basis of a study of stroke patients who experienced blindness and of lesion studies performed on several species, and subsequently championed by Hermann Munk (1881)—and a competing proposal to localize vision in the angular gyrus, defended by David Ferrier (1876) (figure 5.1). Ferrier arrived at his proposal as a result of investigations in which he stimulated the angular gyrus with mild electric shock and caused monkeys to move their eyes to the opposite side, and he further defended it on the basis of experiments in which bilateral lesions of the angular gyrus produced blindness. Ferrier was pretty much alone in advocating the angular gyrus. Neuroanatomical studies by Pierre Gratiolet (1854) and Theodor Meynert (1870) indicated that the optic tract, which first projected to an area of the thalamus known as the *lateral geniculate nucleus* (LGN), then projected on to the occipital lobe (Meynert traced the projections more specifically to the area surrounding the calcarine fissure). Numerous investigators (e.g., Munk 1881; Schäfer 1888b) offered evidence of visual deficits after occipital lobe lesions in animals and clinical reports were amassed from human patients (Wilbrand 1890; Henschen 1893) that traced visual deficits to damage of the occipital lobe. These generally convinced the scientific community, but because of Ferrier's stature the issue was for a time hotly disputed.

During this time, brain researchers were also becoming aware of the distinctiveness of the rear part of the occipital cortex. Already in 1776 Francesco Gennari, in the course of examining frozen sections of human brains, had identified a white stripe that was especially prominent in the posterior part of the brain (Glickstein and Rizzolatti 1984). Subsequently Paul Flechsig (1896) identified this striped area as the target of the projections from the LGN, and Grafton Elliot Smith (1907) named it the *area striata*, a term from which one of its current designations, *striate cortex*, was derived. During this same period, the neuron doctrine was gaining ascendency, and researchers began to map the cortex in terms of neuroanatomical features such as the thickness of particular layers in various parts of the cortex. Of the several major researchers who investigated the cytoarchitecture of mam-

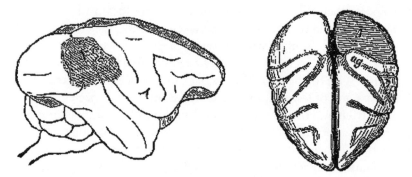

Figure 5.1 Competing proposed localizations of visual processing based on lesions made in monkeys. David Ferrier claimed that lesions to the angular gyrus (left) resulted in blindness and thus that this was the area of visual processing, whereas Hermann Munk argued that lesions to the occipital lobe (right) generated blindness and that it was the locus of visual processing.

malian cortex, the best known today is Korbinian Brodmann (1994 [1909]), who assigned this area the number 17 since it was the seventeenth cortical area he examined (figure 5.2).

As challenging as it sometimes is to establish, identifying a locus of control for a function is only a preliminary step in developing an explanation. A simple identification of a function with a structure does not offer any explanation since it provides no decomposition of that task into more basic operations (Bechtel and Richardson 1993). To invoke information about neural localization to foster development of a decomposition requires either (1) discovering other structures that are involved in carrying out the same function, thus revealing that the first site was not the sole locus of control and provoking the question of what distinct contribution that locus makes, or (2) discovering components within the structure in question, and then asking what activities each of these performed. In the ensuing century of research on visual processing, both of these approaches played a critical role.

Beyond Direct Localization: Complexity within Area 17

A first step beyond simply identifying Brodmann's area 17 as the locus of control had already been taken in a detailed study made by Salomon Henschen of lesion sites that produced vision deficits in humans (Henschen 1893). He showed that deficits in different parts of the occipital lobe produced blindness in different parts of the visual field

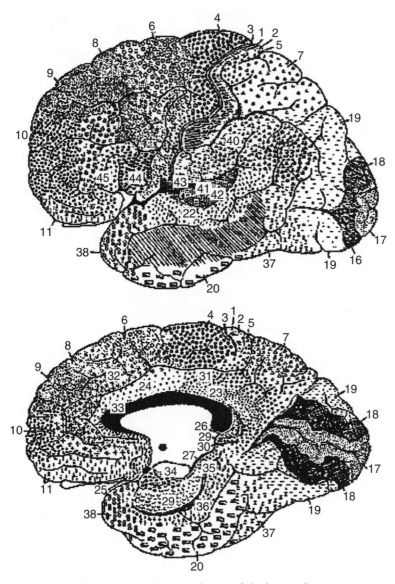

Figure 5.2 Brodmann's cytoarchitectural map of the human brain.

and argued for a topographical order in which different parts of the retina projected onto different areas of the visual cortex (leading him to refer to it as the *cortical retina*). In fact Henschen's map of the visual area in the occipital lobe reversed the organization that was later accepted, a development not uncommon in the history of science and one indicative of how difficult it is to extend beyond individual findings to generate a systematic account. By studying twenty-nine individuals who had sustained damage to the occipital lobe during the Russo-Japanese war,[13] Tatsuji Inouye was able to determine that the central part of the visual field projects to the rear of the occipital lobe and the peripheral parts to the front; a similar study by Gordon Holmes and William Tindall Lister during World War I generated a more detailed and accessible diagram, revealing the topographical projection of parts of the visual field onto the visual cortex.

Lesion studies were limited to demonstrating blindness in a part of the visual field when the relevant part of the visual cortex was removed. The technique that proved most useful in determining just what task area 17 cells were performing was single-cell recording, in which one records the electrical response of a cell to different stimuli. The model for this work was Steven Kuffler's recording from ganglion cells in the retina (Kuffler 1953). Using dark and light circles as stimuli, he discovered that the receptive fields of these cells were organized so that a cell might respond when the center of its receptive field was light but the surrounding area was dark (an *on-center* cell) or the reverse (an *off-center* cell) (figure 5.3, A and B). This protocol was extended to area 17 by two researchers in Kuffler's laboratory at Johns Hopkins, David Hubel and Torsten Wiesel. They began collaborating in 1958 and soon found cells in V1 that responded most vigorously not to spots of light but to oriented lines or bars. The idea that bars of light, not spots, would drive area 17 cells was not obvious and resulted from an accident. Hubel and Wiesel had begun by trying a variety of circular stimuli comparable to those Kuffler had used, but they failed to produce any reliable results. But as they were inserting a glass slide into their projecting ophthalmoscope, Hubel reports that "over the audio-monitor the cell went off like a machine gun" (Hubel 1982, 438). They soon figured out that it was not the dot on the slide that was producing the response, but the fact that "as the glass slide was inserted its edge was casting onto the retina a faint but sharp shadow, a straight dark line on a light background" (439).[14]

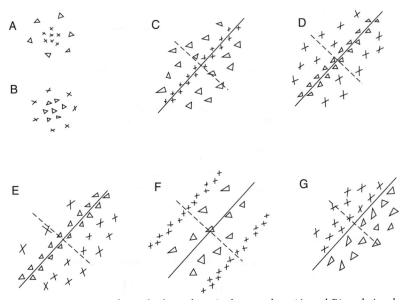

Figure 5.3 Responses from the lateral geniculate nucleus (A and B) and simple cortical cells in area 17 (C–G) in the cat. Xs indicate areas producing excitatory responses and △s areas producing inhibitory responses. (From Hubel and Wiesel 1962.)

Over their first ten years of collaboration, Hubel and Wiesel probed the striate cortex of both cats (Hubel and Wiesel 1962) and monkeys (Hubel and Wiesel 1968) in a manner that suggested a decomposition of function within area 17. Cells differed in their response properties. What they termed *simple cells* had receptive fields with spatially distinct *on* and *off* areas along a line at a particular orientation, which varied from cell to cell (figure 5.3). Hubel and Wiesel proposed how several cells with center-surround receptive fields (such as those found in the LGN) whose center lay on the line might send excitatory input to a receptor cell that would serve as an AND-gate and fire if all of its input cells were active. In this regard, it is salient that simple cells predominate in layer 4, the layer of cortical cells that receives inputs. Whereas simple cells were sensitive to stimuli only at a given retinal location, what Hubel and Wiesel termed *complex cells* were responsive to bars of light at a particular orientation anywhere within their receptive fields (figure 5.4). Many complex cells were also sensitive to the direction of movement of bars within their receptive fields. Hubel

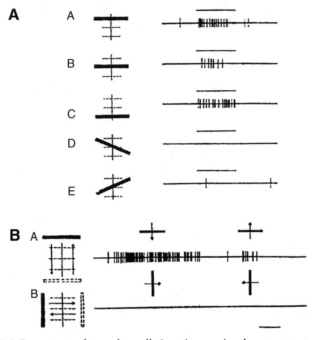

Figure 5.4 Responses of complex cells in primary visual cortex to stationary, horizontally oriented black rectangles at different locations in the cell's receptive field (A) or moving through the receptive field in a specific direction (B). (From Hubel and Wiesel 1962.)

and Wiesel identified these as *complex cells* since their response pattern could be explained if they received input from several simple cells, any of which would be sufficient to cause the complex cell to fire. Complex cells are found primarily in layers 2 and 3 and 4 and 6.[15] In their papers from this period Hubel and Wiesel also distinguished *hypercomplex cells*, which responded maximally only to bars extending just the width of their receptive fields. They proposed that the successive generation of particular responses from these different cell types was the major contribution of area 17:

The elaboration of simple cortical fields from geniculate concentric fields, complex from simple, and hypercomplex from complex is probably the prime function of the striate cortex—unless there are still other as yet unidentified cells there. One need not assume, of course, that the output consists entirely of the axons of the hypercomplex cells, the other types being merely interposed as links between input and output. We know, for example, that in the cat the posterior corpus callosum contains axons of all three cell types. (Hubel and Wiesel 1968, 239)

Discovering the orientation sensitivity and the distinction between simple, complex, and hypercomplex cells provided an initial suggestion as to the functions actually performed in area 17.[16] But even more important, their research demonstrated that area 17 could not be the sole locus of visual processing, since seeing is not just detecting bars of light at particular orientations. Hubel and Wiesel conclude their 1968 paper with the prophetic comment:

Specialized as the cells of 17 are, compared with rods and cones, they must, nevertheless, still represent a very elementary stage in the handling of complex forms, occupied as they are with a relatively simple region-by-region analysis of retinal contours. How this information is used at later stages in the visual path is far from clear, and represents one of the most tantalizing problems for the future. (Hubel and Wiesel 1968, 242)

Beyond Direct Localization: Identifying Prestriate Visual Areas

The second of the two means identified previously for moving beyond simply localizing visual processing in area 17 was discovering other areas of adjoining cortex that are involved in visual processing.[17] Through the first half of the twentieth century (during which time the brain sciences were dominated by a strong anti-localizationist sentiment), areas in front of the primary visual cortex were widely assumed to be association areas in which individual visual images were associated with other sensory input and memories of previous stimuli so as to identify the objects and events seen. Visual processing per se was thought to be limited to area 17. Accordingly Karl Lashley (1950, 467), who had himself coined the term *prestriate region* for the area surrounding the striate cortex, wrote: "visual habits are dependent upon the striate cortex and upon no other part of the cerebral cortex."

For many researchers, one sign of the lack of differentiated function beyond area 17 was that these areas did not seem to be topographically organized in the manner of striate cortex. Accordingly one of the first positive indications of another visual processing area was Alan Cowey's (1964) discovery, using surface electrodes to record evoked responses, of a second topographically organized area in Brodmann's area 18, which immediately adjoins area 17. Using single-cell recording with implanted electrodes, Hubel and Wiesel (1965) confirmed the topographical organization of this area and identified yet a third such area in Brodmann's area 19. As these additional areas were identified, area 17 came to be known as *primary visual cortex* or V1, the area discovered by Cowey as V2, and the new area found by Hubel and Wiesel as V3. By tracing degeneration of fibers from discrete lesions in striate cortex to areas in surrounding cortex, Semir Zeki (1969) offered supporting evidence for the demarcation of these two topologically organized areas. He then extended this approach by creating lesions in V2 and V3, from which he traced degeneration forward into areas on the anterior bank of the lunate sulcus in which "the organized topographic projection, as anatomically determined, gradually breaks down" (Zeki 1971, 33).[18] Zeki labeled these areas V4 and V4a.[19]

For the discovery of these additional areas to advance the functional decomposition of vision, it was necessary to link different areas with distinct functions. As with V1, single-cell recording played the major role. Zeki (1973) recorded from cells in V4 and found "in every case the units have been colour coded, responding vigorously to one wavelength and grudgingly, or not at all, to other wavelengths or to white light at different intensities" (422).[20] Zeki (1974) also recorded from cells on the posterior bank of the superior temporal sulcus (which he labeled V5 but others refer to as MT), which responded to movement.[21]

Beyond Direct Localization: Expanding Visual Analysis into Temporal and Parietal Cortexes

The discovery in extrastriate cortex of visual processing areas that analyzed distinct visual properties such as color and motion both significantly advanced the functional decomposition of vision and posed a major question: where is the information about edges, colors, and motion put to use to permit the recognition of objects and events in the world? The first suggestions that areas in the temporal lobe might play such a role were put forth in a study by Edward Schäfer (1888a) ostensibly devoted to showing that, contrary to Ferrier's claims, the temporal cortex was not the locus of an auditory center. In monkeys in which either the superior temporal gyrus or nearly all the temporal lobes were removed (figure 5.5), Schäfer reported no detectable loss of hearing but described a deficit in recognizing visually presented stimuli:

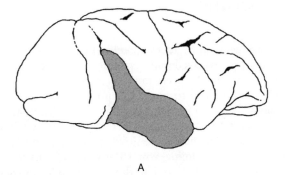

A

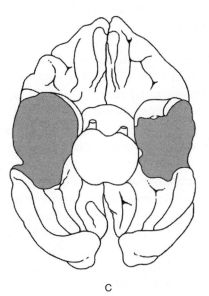

B

C

Figure 5.5 Schäfer's diagrams of areas in temporal cortex in which lesions in monkeys resulted in the monkey's inability to understand what it saw.

The condition was marked by loss of intelligence and memory, so that the animals, although they received and responded to impressions from all the senses, appeared to understand very imperfectly the meaning of such impressions. This was not confined to any one sense, and was most evident with visual impressions. For even objects most familiar to the animals were carefully examined, felt, smelt and tasted exactly as a monkey will examine an entirely strange object, but much more slowly and deliberately. And on again, after only a few minutes, coming across the same object, exactly the same process of examination would be renewed, as if no recollection of it remained. (Schäfer 1888a, 375)

Little attention was paid to Schäfer's observations until a study by Heinrich Klüver and Paul Bucy in the late 1930s in which removal of the temporal lobe in monkeys resulted in a condition they described as *psychic blindness* or *visual agnosia,* in which "the ability to recognize and detect the meaning of objects on visual criteria alone seems to be lost although the animal exhibits no or at least no gross defects in the ability to discriminate visually" (Klüver 1948, 151). In Klüver and Bucy's studies the monkeys also exhibited a variety of other behavioral changes, including loss of emotional responsiveness and increased sexual behavior, but Karl Pribram and Muriel Bagshaw (1953) demonstrated that the different effects resulted from different lesions in temporal cortex. Visual agnosia, in particular, followed lesions of the amygdala and adjacent cortex. Subsequently Pribram collaborated with Mortimer Mishkin in localizing visual agnosia specifically to lesions in inferotemporal cortex (Mishkin and Pribram 1954). Later Mishkin (1966) demonstrated that TE and TEO, areas within inferotemporal cortex that von Bonin and Bailey (1951) had distinguished on cytoarchitectonic grounds, produced differential deficits, with TEO lesions producing greater deficits in single-pattern discrimination tasks and TE lesions generating greater deficits on learning to perform multiple discriminations in parallel.

Again the lesion studies indicating separate processing areas were complemented by single-cell recording studies that sought to determine which stimuli generated specific responses in inferotemporal cortex. Gross et al. (1972) found cells in the inferotemporal cortex of the macaque that responded most vigorously to shapes such as hands. (Reminiscent of the serendipity of Hubel and Wiesel, they discovered the responsiveness to hands quite by accident when, after failing to find a light source that would drive a particular cell, they waved a hand in front of the stimulus screen and produced a vigorous response.) The combination of lesion and single-cell recording studies thus succeeded in pointing to a critical role of inferotemporal cortex in the higher-level visual process of recognizing objects. A similar pattern of first lesion studies (starting with Ferrier and Yeo 1884) and then single-cell recording studies (starting with Hyvärinen and Poranen 1974) figured in research on the parietal cortex, indicating a role for parietal cortex in analyzing the location of stimuli and guiding movements toward them.

Proposing a Complex, Organized System

The research described in the previous two sections clearly advanced the efforts to functionally decompose and localize visual processing. Whereas initially only V1 seemed to be involved, it began to appear as if much of the rear of the brain was devoted to analyzing visual inputs. But at this stage there was no unifying principle with which to characterize how the brain executed visual processing. It was in this context that Mortimer Mishkin and Leslie Ungerleider suggested an overall schema as to how visual processes were organized in the brain (figure 5.6; Mishkin et al. 1983). In their model, visual processing beyond V1 is organized into two pathways, one progressing dorsally into posterior parietal cortex that analyzes *where* objects are in the visual field, the other progressing ventrally down into inferotemporal cortex that analyzes *which* objects are present in the visual scene.[22] This schema quickly attracted a great deal of attention as a way of conceptualizing the decomposition of visual processes, and it has inspired both purely behavioral research (Chen et al. 2000) and computational modeling studies in cognitive science (Rueckl et al. 1989; Jacobs et al. 1991).

Shortly after Mishkin and Ungerleider made their proposal, Livingstone and Hubel (1984) argued for extending the two pathways back into V1, LGN, and the retina, resulting in a model of two processing streams from the very earliest sensory input. One basis for this proposal was the earlier discovery of two different cell types in the retina and the LGN. In primates, the two cell types are referred to as P-ganglion or parvo cells and M-ganglion or magno cells. The parvo cells have small receptive fields, making them sensitive to high spatial frequencies, medium conductance velocities, color, and responses lasting as long as a stimulus is present. In contrast, the magno cells have large receptive fields, rapid conductance velocities, no sensitivity to color,

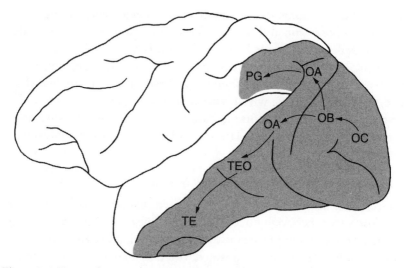

Figure 5.6 Two pathways of visual processing in the rhesus monkey proposed by Mishkin et al. (1983). Each begins in area OC (primary visual cortex, also called V1) and projects into prestriate areas OB (V2) and OA (V3, V4, and MT). The *what* pathway then projects ventrally into inferior temporal cortex (areas TEO and TE), whereas the *where* pathway projects dorsally into inferior parietal cortex (area PG).

and a more transient response, suggesting that they are sensitive to motion. Research on Old World monkeys revealed that this scheme is maintained in the LGN, where the inner two of six cell layers consist of magno cells while the outer four consist of parvo cells.

The challenge was how to link this precortical distinction with Mishkin and Ungerleider's two cortical pathways given that Hubel and Wiesel's early research had not suggested any similar differentiation in V1. A new technique pioneered by Margaret Wong-Riley, involving application of cytochrome oxidase stains to striate cortex, provided suggestive evidence. Cytochrome oxidase is an enzyme critical to the oxidative metabolism of the cell; staining for it reveals areas of high metabolic activity. In layers 2 and 3 and 5 and 6 of V1, cytochrome oxidase stain revealed "blobs"[23]—regions of increased metabolic activity. Recording separately from cells in the blob regions and in the interblob regions, Livingstone and Hubel (1984) found orientation-selective cells only in the interblob regions and wavelength-sensitive cells in the blobs, indicating a separation of processing within V1. The wavelength sensitivity of cells in the blob regions suggested that they represented a continuation of the parvocellular stream and projected onto V4 and the ventral steam. This is supported by the anatomy: The magnocellular layers of the LGN project to layer 4B in V1, where there are no blobs, whereas the parvocellular layers of the LGN project, via layers 4A and 4Cb, to layers 2 and 3 of V1, where there are both blob and interblob regions. Cytochrome oxidase stain also revealed a differentiation in V2 of alternating thick and thin stripes with inter-stripe areas between them. The differentiation in V1 is maintained, with the thick-stripe regions receiving their input from layer 4B, the thin-stripe regions from the blobs of layers 2 and 3, and the interstripe regions from the interblob regions in V1. On the basis of these discoveries, Livingstone and Hubel proposed extending Mishkin and Un-gerleider's two pathways to account for all visual processing.

Grand organizing schemes often play an important role in suggesting that a line of inquiry has obtained a relatively high level of maturity. But equally importantly they often inspire new research designed to show that the system is more complex and perhaps organized along different lines. One shortcoming of Mishkin and Ungerleider's scheme, and especially the extension by Livingstone and Hubel, was that it failed to do justice to the integration between the posited streams. Van Essen and Gallant (1994) point out that processing in later cortical parts of one pathway can continue even when its supposed precortical input is removed, indicating that there is extensive crossover between the two streams early on. They also identify several points within the cortex where there are projections between the two pathways (e.g., between MT and V4). Thus, although their schematic representation (figure 5.7) distinguishes different routes of processing, they prefer to speak of different streams, not pathways. Moreover, the characterization of the two streams as processing "what" and "where" information has been questioned. Milner and Goodale (1995) argue that the dorsal stream involving the parietal lobe (supposedly only involved in processing location information) receives information about the identity of objects (revealed in the ability of individuals with temporal lobe lesions to be able to grasp objects appropriately for their function). They propose that what is distinctive about parietal processing is that it is primarily concerned with coordinating information about visual stimuli for action, whereas the ventral stream is principally involved in extracting information about visual stimuli required for higher cognitive processing.

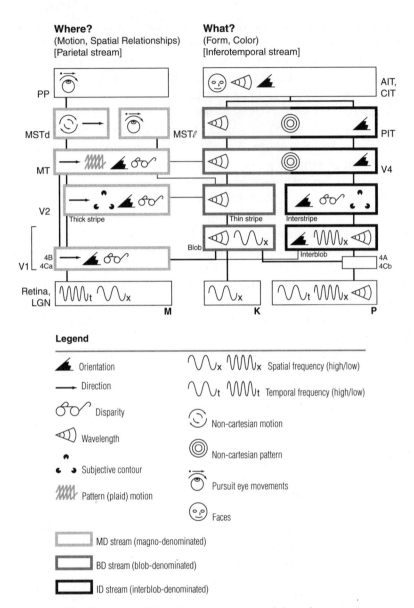

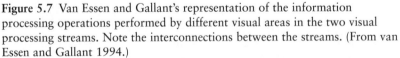

Figure 5.7 Van Essen and Gallant's representation of the information processing operations performed by different visual areas in the two visual processing streams. Note the interconnections between the streams. (From van Essen and Gallant 1994.)

Even with such qualifications, though, the decomposition into different processing areas and their organization into streams, as indicated in van Essen and Gallant's figure, has gained substantial general support, not only in neuroscience but also in psychology and the cognitive sciences generally. Visual processing is recognized to be the product of a complex machine. Future research will undoubtedly help revise the current decomposition. But this line of research has reached a level of maturity that serves as an exemplar, in Kuhn's sense, for other domains of cognitive neuroscience.

Conclusion

Although twenty years ago the prospects may have seemed dim for truly bridging the gap between cognition and the brain, today cognitive neuroscience is a thriving endeavor. The success in decomposing visual processing has inspired hope that it will prove possible to discover what functions component systems in the brain perform and to develop models of how, through the coordinated performance of these functions, the brain enables us to live the cognitive lives we do. Moreover, as these successes multiply, old philosophical worries about connecting mind and brain—worries based on multiple realizability and on the idea that at most empirical research could show a correlation between brain function and mental function—have largely withered away (Bechtel and McCauley 1999). A principal reason for this is that researchers adopted the idea that mental functions are performed by different brain parts as a working hypothesis and used a combination of behavioral and neural tools to decompose progressively the processes of vision. Once a schema for a mechanism involving parts of the brain performing different cognitive functions is developed, worries about the mind-body problem seem very much beside the point.

NOTES

1. As with most integrations in science, this one has incorporated only parts of the contributing fields—primarily cognitive psychology on the cognitive side and systems and behavioral neuroscience on the neural side. Although computational modeling, such as is pursued in artificial intelligence, has played a role in cognitive neuroscience (mostly in the form of neural network models that seek to include more neural detail than many connectionist models), it has primarily been pursued

in a separate cross-disciplinary enterprise referred to as *computational neuro-science*.

2. PET injects into the blood a radioactive tracer such as ^{15}O, which, on decay, releases a positron. After traveling a very short distance this particle usually encounters an electron and splits into two gamma rays or photons, which follow a trajectory 180° apart; the scanner is designed to detect simultaneously arriving photons, and the tomographic program is able to compute the initial site of gamma ray formation. Generally a coloring scheme is used to indicate the degree of radioactivity present at various locations. Increased blood flow, which is assumed to relate to increased neural activity, results in increased positron emission; accordingly, researchers construe areas of increased positron activity as areas of increased neural activity and refer to increased activation in the area.

3. The term *neural imaging* often refers as well to other techniques, such as evoked response potential investigations employing electroencephalography (endogenous electrical signals recordable on the scalp, a technique that has been available for several decades but that was much refined in the 1990s), magneto-encephalography, and optical imaging.

4. Miller and Gazzaniga had been collaborators in research supported by one of the earlier Sloan Foundation grants, one of the few grants during the early period to have a strong focus on the brain. Their goal was to analyze deficits stemming from brain damage so as to draw inferences about cognitive function in non-brain-damaged individuals. In 1981 Gazzaniga established a Cognitive Neuro-science Institute, which received a Sloan grant of $500,000.

5. Another goal of the study panels was to publish a book containing reports from each panel, for which arrangements had been made with MIT. This book was intended to provide a perspective on the current state of each of the selected fields. However, only some of the study panels issued reports, and they were never edited into such a book.

6. This center was later transferred to the University of California, Davis, when Gazzaniga moved there.

7. The support went beyond the universities at which centers were established. The two foundations also supported a broad range of postdoctoral training and individual research grants to investigators at institutions without center grants.

8. It is important to note the distinction between reductive materialism and eliminative materialism. Reductive materialism claims that cognitive processes will turn out to be identical to brain processes, and thus that they will be retained in the final ontology at which science arrives, while eliminative materialism proposes that cognitive processes will not map onto brain processes and will accordingly be replaced by newly characterized processes derived from understanding the brain processes. Both, however, put primacy on the neural foundations—either to vindicate the cognitive taxonomy or to supplant it.

9. The metaphor also mischaracterizes computer science: hardware is not designed independently of concern for function, the software that runs on it. Rather, the functional characteristics of hardware are of primary concern and affect how any piece of software written for it will function.

10. Brodmann's goal was to differentiate brain areas that were functionally important, and he appealed to criteria such as relative thickness of cortical layers because he thought they were likely to be of functional significance. Patterns of connectivity, of obvious functional significance, have been one of the major tools for identifying brain areas. The development of tracers to reveal the targets of axons was one of the major advances in neuroanatomy in the 1970s.

11. For most philosophers, the decomposition bottoms out in propositional attitudes, but for most cognitive scientists, the fundamental operations within the cognitive machine are lower-level information processing activities.

12. The approach, though, may capture all a scientist requires out of a reduction, in that, for the behavior of any component of interest, one may ask what subcomponents of that component, in the confines of a system of a particular kind, enable it to behave as it does.

13. What was critical for developing such maps was that the patients had suffered focal damage. As Glickstein (1988) noted, one explanation for their restricted damage was the introduction of new bullets by the Russians that often caused serious injury but not death.

14. Hubel describes some of the sense of surprise at the finding that individual cells responded to a bar of light at a particular orientation: "This was unheard of. It is hard, now, to think back and realize just how free we were from any idea of what cortical cells might be doing in an animal's daily life" (1982, 439).

15. An important difference that helps to distinguish the function of the different layers is that they generally project to different brain areas: layers 2 and 3 to other cortical areas; layer 5 to the superior colliculus, pons, and pulvinar; and layer 6 back to the LGN.

16. Hubel and Wiesel's research of the time also revealed other features of the organization of area 17: that there were alternating columns of cells responsive to each eye and that within these columns adjacent cells seemed to respond to bars with gradually changing orientations.

17. By analyzing patients with cortical achromatopsia (the inability to see colors) whose lesions could be traced to the fusiform gyrus adjacent to the striate cortex, both Verrey (1888) and MacKay and Dunlop (1899) had provided evidence of a second visual area, one devoted to color perception. However, most late nineteenth- and early twentieth-century researchers dismissed these claims in favor of the supposition of one cortical center for vision in the striate cortex, which might produce achromatopsia with mild lesions and full blindness with more serious lesions (Zeki 1993). One finding supporting this interpretation was that most cases of achromatopsia also manifested scotomas or areas of total blindness, suggesting that one lesion produced both effects.

18. Zeki (1971, 34) ends the paper with the following comment about projections to other brain areas: "How the prestriate cortex is organized in regions beyond (central to) V4 and V4a remains to be seen. It is perhaps sufficient to point out at present that the organisation of the prestriate areas would seem to be far more complicated than previously envisaged and that the simplistic wiring diagram from area 17 to area 18, from area 18 to area 19 and from area 19 to the so-

called 'inferior temporal' area will have to be abandoned. At any rate, we were not able in this study to find any projections to the 'inferior temporal' areas from areas 18 and 19 (V2 and V3)."

19. During the same period John Allman and Jon Kaas (1971), through single-cell recording in squirrel monkeys, traced topographically organized visual areas not only into extrastriate regions but into temporal and parietal cortexes.

20. Zeki's discovery of a color area made explicable the clinical reports of specific deficits in color perception (achromatopsia) described in note 17. These patients had presumably suffered lesions in V4.

21. In 1983 Zihl et al. reported on a patient who, as a result of vascular damage, could not perceive motion; to the patient, activities such as coffee being poured into a cup appeared as contiguous shapes, like a glacier. Zeki's discovery of motion detection by V5 or MT could likewise explain this patient's deficit as due to damage in that area.

22. The distinction between *what* and *where* processing had been advanced previously by Schneider (1967) and Trevarthen (1968) for subcortical areas, but by proposing it for cortical areas Mishkin and Ungerleider offered a macro-level organizing principle for thinking about cortical visual areas that integrated the findings of visual processing in both extrastriate areas.

23. Livingstone and Hubel (1984) introduced the term *blobs* to characterize their appearance, citing the *Oxford English Dictionary* for the term. These blobs are "oval, measure roughly 150 × 200 μm, and in the macaque monkey lie centered along ocular dominance columns, to which their long axes are parallel" (310).

REFERENCES

Allman, J. M., and J. H. Kaas. 1971. "A Representation of the Visual Field in the Caudal Third of the Middle Temporal Gyrus of the Owl Monkey *Aotus trivirgatus.*" *Brain Research* 31: 85–105.

Bechtel, W., and R. N. McCauley. 1999. "Heuristic Identity Theory or Back to the Future: The Mind-Body Problem against the Background of Research Strategies in Cognitive Neuroscience." In M. Hahn and S. C. Stoness, eds., *Proceedings of the 21st Meeting of the Cognitive Science Society.* Mahwah, N.J.: Lawrence Erlbaum, 67–72.

Bechtel, W., and J. Mundale. 1999. "Multiple Realizability Revisited: Linking Cognitive and Neural States." *Philosophy of Science* 66: 175–207.

Bechtel, W., and R. C. Richardson. 1993. *Discovering Complexity: Decomposition and Localization as Scientific Research Strategies.* Princeton, N.J.: Princeton University Press.

Bechtel, W., A. Abrahamsen, and G. Graham. 1998. "The Life of Cognitive Science." In W. Bechtel and G. Graham, eds., *A Companion to Cognitive Science.* Oxford: Basil Blackwell, 1–104.

Brodmann, K. 1994 [1909]. *Localisation in the Cerebral Cortex*, transl. L. J. Garey. London: Smith-Gordon.

Chen, J., J. Myerson, S. Hale, and A. Simon. 2000. "Behavioral Evidence for Brain-Based Ability Factors in Visuospatial Information Processing." *Neuropsychologia* 38: 380–87.

Cowey, A. 1964. "Projection of the Retina on to Striate and Prestriate Cortex in the Squirrel Monkey *Saimiri sciureus*." *Journal of Neurophysiology* 27: 366–93.

Dennett, D. C. 1971. "Intentional Systems." *Journal of Philosophy* 68: 87–106.

Elliot Smith, G. 1907. "New Studies on the Folding of the Visual Cortex and the Significance of the Occipital Sulci in the Human Brain." *Journal of Anatomy* 41: 198–207.

Ferrier, D. 1876. *The Functions of the Brain*. London: Smith, Elder.

Ferrier, D., and G. F. Yeo. 1884. "A Record of the Experiments on the Effects of Lesions of Different Regions of the Cerebral Hemispheres." *Philosophical Transactions of the Royal Society of London* 175: 479–564.

Flechsig, P. E. 1896. *Gehirn und Seele*. Leipzig: Veit.

Fodor, J. A. 1974. "Special Sciences or: The Disunity of Science as a Working Hypothesis." *Synthese* 28: 97–115.

Fox, P. T., M. A. Minton, M. E. Raichle, F. M. Miezin, J. M. Allman, and D. C. van Essen. 1986. "Mapping Human Visual Cortex with Positron Emission Tomography." *Nature* 323: 806–9.

Glickstein, M. 1988. "The Discovery of the Visual Cortex." *Scientific American* 259(3): 118–27.

Glickstein, M., and G. Rizzolatti. 1984. "Francesco Gennari and the Structure of the Cerebral Cortex." *Trends in Neuroscience* 7: 464–67.

Gratiolet, P. 1854. "Note sur les Expansions des Racines Cérébrales du Nerf Optique et sur leur Terminaison dans une Région Déterminée de l'Écorce des Hémisphères." *Comptes Rendus Hebdomadaires des Séances de l'Académie des Sciences de Paris* 29: 274–78.

Gross, C. G., C. E. Rocha-Miranda, and D. B. Bender. 1972. "Visual Properties of Neurons in Inferotemporal Cortex of the Macaque." *Journal of Neurophysiology* 35: 96–111.

Henschen, S. E. 1893. "On the Visual Path and Centre." *Brain* 16: 170–80.

Hubel, D. H. 1982. "Evolution of Ideas on the Primary Visual Cortex, 1955–1978: A Biased Historical Account." *Bioscience Reports* 2: 435–69.

Hubel, D. H., and T. N. Wiesel. 1962. "Receptive Fields, Binocular Interaction and Functional Architecture in the Cat's Visual Cortex." *Journal of Physiology* (London) 160: 106–54.

——. 1965. "Receptive Fields and Functional Architecture in Two Non-Striate Visual Areas (18 and 19) of the Cat." *Journal of Neurophysiology* 195: 229–89.

——. 1968. "Receptive Fields and Functional Architecture of Monkey Striate Cortex." *Journal of Physiology* (London) 195: 215–43.

Hyvärinen, J., and A. Poranen. 1974. "Function of the Parietal Associative Area as Revealed from Cellular Discharges in Alert Monkeys." *Brain* 97: 673–92.

Jacobs, R. A., M. I. Jordan, and A. G. Barto. 1991. "Task Decomposition through Competition in a Modular Connectionist Architecture: The What and Where Vision Tasks." *Cognitive Science* 15: 219–50.

Klüver, H. 1948. "Functional Differences between the Occipital and Temporal Lobes with Special Reference to the Interrelations of Behavior and Extra-cerebral Mechanisms." In L. Jeffress, ed., *Cerebral Mechanisms in Behavior.* New York: Wiley, 147–99.

Kuffler, S. W. 1953. "Discharge Patterns and Functional Organization of Mammalian Retina." *Journal of Neurophysiology* 16: 37–68.

Lashley, K. S. 1950. "In Search of an Engram." *Symposia of the Society for Experimental Biology* 4: 454–80.

LeDoux, J. E., and W. Hirst. 1986. *Mind and Brain: Dialogues in Cognitive Neuroscience.* Cambridge: Cambridge University Press.

Livingstone, M. S., and D. H. Hubel. 1984. "Anatomy and Physiology of a Color System in the Primate Visual Cortex." *Journal of Neuroscience* 4: 309–56.

Machamer, P., L. Darden, and C. F. Craver. 2000. "Thinking about Mechanisms." *Philosophy of Science* 67: 1–25.

MacKay, G., and J. C. Dunlop. 1899. "The Cerebral Lesions in a Case of Complete Colour Blindness." *Scottish Medical and Surgical Journal* 5: 503–12.

Meynert, T. 1870. "Beiträge zur Kenntniss der centralen Projection der Sinnesoberflächen." *Sitzungsberichte der Kaiserlichen Akademie der Wissenschaften, Wien. Mathematisch- Naturwissenschaftliche Classe* 60: 547–62.

Milner, A. D., and M. G. Goodale. 1995. *The Visual Brain in Action.* Oxford: Oxford University Press.

Mishkin, M. 1966. "Visual Mechanisms beyond the Striate Cortex." In R. W. Russel, ed., *Frontiers in Physiological Psychology.* New York: Academic Press, 93–119.

Mishkin, M., and K. Pribram. 1954. "Visual Discriminative Performance following Partial Ablations of the Temporal Lobe. I. Ventral vs. Lateral." *Journal of Comparative and Physiological Psychology* 47: 14–20.

Mishkin, M., L. G. Ungerleider, and K. A. Macko. 1983. "Object Vision and Spatial Vision: Two Cortical Pathways." *Trends in Neurosciences* 6: 414–17.

Munk, H. 1881. *Über die Funktionen der Grosshirnrinde.* Berlin: A. Hirschwald.

Panizza, B. 1855. "Osservazioni sul Nervo Ottico." *Memoria, Instituo Lombardo di Scienze, Lettere e Arte* 5: 375–90.

Petersen, S. E., P. J. Fox, M. I. Posner, M. Mintun, and M. E. Raichle. 1989. "Positron Emission Tomographic Studies of the Processing of Single Words." *Journal of Cognitive Neuroscience* 1(2): 153–70.

Posner, M. I. 1978. *Chronometric Explorations of Mind.* Hillsdale, N.J.: Lawrence Erlbaum.

Posner, M. I., and P. McLeod. 1982. "Information Processing Models: In Search of Elementary Operations." *Annual Review of Psychology* 33: 477–514.

Pribram, K. H., and M. Bagshaw. 1953. "Further Analysis of the Temporal Lobe Syndrome Utilizing Fronto-Temporal Ablations." *Journal of Comparative Neurology* 99: 347–75.

Putnam, H. 1967. "Psychological Predicates." In W. H. Capitan and D. D. Merrill, eds., *Art, Mind and Religion*. Pittsburgh: University of Pittsburgh Press, 37–48.

Rueckl, J. G., K. R. Cave, and S. M. Kosslyn. 1989. "Why are 'What' and 'Where' Processed by Separate Cortical Visual Systems? A Computational Investigation." *Journal of Cognitive Neuroscience* 1: 171–86.

Schäfer, E. A. 1888a. "Experiments on Special Sense Localisations in the Cortex Cerebri of the Monkey." *Brain* 10: 362–80.

——. 1888b. "On the Functions of the Temporal and Occipital Lobes: A Reply to Dr. Ferrier." *Brain* 11: 145–66.

Schneider, G. E. 1967. "Contrasting Visuomotor Functions of Tectum and Cortex in the Golden Hamster." *Psychologische Forschung* 31: 52–62.

Sternberg, S. 1969. "The Discovery of Processing Stages: Extension of Donders' Method." *Acta Psychologica* 30: 276–315.

Trevarthen, C. 1968. "Two Mechanisms of Vision in Primates." *Psychologische Forschung* 31: 299–337.

Van Essen, D. C., and J. L. Gallant. 1994. "Neural Mechanisms of Form and Motion Processing in the Primate Visual System." *Neuron* 13: 1–10.

Verrey, D. 1888. "Hémiachromatopsie droite absolue." *Archives d'Ophthalmologie* 8: 289–301.

Von Bonin, G., and P. Bailey. 1951. *The Isocortex of Man*. Urbana: University of Illinois Press.

Wilbrand, H. 1890. *Die hemianogischen Gesichtsfeld-Formen und das optische Wahrnehmungszentrum*. Wiesbaden: J. F. Bergmann.

Zeki, S. M. 1969. "Representation of Central Visual Fields in Prestriate Cortex Monkey." *Brain Research* 14: 271–91.

——. 1971. "Cortical Projections from Two Prestriate Areas in the Monkey." *Brain Research* 34: 19–35.

——. 1973. "Colour Coding of the Rhesus Monkey Prestriate Cortex." *Brain Research* 53: 422–27.

——. 1974. "Functional Organization of a Visual Area in the Posterior Bank of the Superior Temporal Sulcus of the Rhesus Monkey." *Journal of Physiology* 236: 549–73.

——. 1993. *A Vision of the Brain*. Oxford: Blackwell.

Zihl, J., D. von Cramon, and N. Mai. 1983. "Selective Disturbance of Movement Vision after Bilateral Brain Damage." *Brain* 106: 313–40.

6

Discovering Mechanisms in Neurobiology

The Case of Spatial Memory

Carl F. Craver
Department of Philosophy, Florida International University, Miami, Florida

Lindley Darden
Committee on the History and Philosophy of Science, University of Maryland, College Park, Maryland

This chapter is about discovery in neurobiology; more specifically, it is about the discovery of mechanisms. The search for mechanisms is widespread in contemporary neurobiology, and, understandably, the character of this product shapes the process by which mechanisms are discovered. Analyzing mechanisms and their characteristic organization reveals constraints on their discovery. These constraints reflect, at least in part, what it is to have a plausible description of a mechanism. These constraints also highlight varieties of evidence that both guide and delimit the construction, evaluation, and revision of such plausible descriptions.

The central example in the following discussion is the continuing discovery of the mechanism of spatial memory. Spatial memory, roughly speaking, is the ability to learn to navigate through a novel environment. The mechanism of spatial memory is multilevel, and recently an integrated sketch of the mechanism at each of these levels has started to emerge. Even though this sketch is far from complete at this time, the example offers a glimpse at the kinds of constraints that are delimiting and guiding this gradual and piecemeal discovery process.

This chapter opens with an analysis of mechanism, discussing their components and their characteristic spatial, temporal, and multilevel organization. Mechanisms are often discovered gradually and piecemeal. The second section introduces conventions for constructing incomplete and abstract descriptions of mechanisms (namely the mechanism sketch and the mechanism schema) and for describing the constraints under which the gradual and piecemeal discovery of these sketches and schemata proceeds. The third section uses the case study of spatial memory to illustrate how constraints on the organization of mechanisms have guided and delimited their discovery. The final section focuses on hierarchical constraints in particular and discusses the use of multilevel experiments to integrate the different levels in such a description. Throughout, the goal is to show that the products, multilevel descriptions of mechanisms, shape the process by which they are discovered.

Mechanisms and Their Organization

Mechanisms

Neurobiologists often speak of "systems" and "cascades" to describe what we call, also consistently with the field's language, "mechanisms." Through our collaboration with Peter Machamer, we have come to think about mechanisms as follows. Mechanisms are collections of entities and activities organized in the production of regular changes from start or setup conditions to finish or termination conditions (Machamer et al. 2000). Entities in neurobiology include such things as pyramidal cells, neurotransmitters, brain regions, and mice. Activities are the various doings in which these entities engage: pyramidal cells *fire*, neurotransmitters *bind*, brain regions *process*, and mice *swim* in water while eagerly *searching* for a means of escape. When neurobiologists speak generally about activities, they use a variety of terms; activities are often called "processes," "functions," and "interactions." Activities are the things that entities do; they are the productive components of mechanisms, and they constitute the stages of mechanisms.

Organization

The entities and activities composing mechanisms are *organized*; they are organized such that they *do* something, *carry out* some process, *exercise* some faculty, *perform* some function, or *produce* some end product. We refer to this activity or behavior of the mechanism as a whole as the *phenomenon* to be explained by the description of the

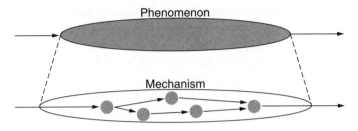

Figure 6.1 Phenomenon and mechanism.

mechanism. This is the activity at the top of figure 6.1. Below it are the entities and activities composing the mechanism for that phenomenon.

The phenomena to be explained by descriptions of mechanisms can be understood in the spirit of Bogen and Woodward (1988, 317). We think of phenomena as relatively stable and repeatable properties or activities that can be produced, manipulated, or detected in a variety of experimental arrangements. Examples of phenomena in neurobiology include the acquisition, storage, and retrieval of spatial memories; the release of neurotransmitters; and the generation of action potentials.

Mechanisms are *organized* in the production of phenomena. One aspect of this organization is *temporal*. The stages of mechanisms have a productive order from beginning to end, with earlier stages giving rise to later stages. The stages of mechanisms also have characteristic rates and durations that can be crucial to their operation. Order, rate, and duration are crucial, for example, for the generation of action potentials in neurons: sodium channels open before potassium channels, and the respective timing and duration of their opening account for the characteristic waveform of the action potential.

A second aspect of the organization of mechanisms is *spatial*. Different stages of the mechanism may be *compartmentalized* within some boundary or otherwise *localized* within some more or less well-defined region. These stages are *connected* with one another by, for example, motion and contact. Often, the connection between stages depends crucially upon the *structures* of the entities in the mechanism and upon those structured entities being *oriented* with respect to one another in particular ways.

These temporal and spatial aspects of the organization of mechanisms trace the productive relationships among the component stages—the relationship of one stage giving rise to, driving, making, or allowing its successor. Importantly, mechanisms exhibit a productive continuity, without gaps, from setup to termination. Mechanisms require productive continuity to work, and accordingly our understanding of mechanisms turns on our ability to establish a seamless continuity between setup and termination. The discovery of mechanisms is often driven by the goal of eliminating gaps in this productive continuity.

Consider an example from contemporary neurobiology, the mechanism of long-term potentiation (LTP). LTP is a means of strengthening synapses, and many think that LTP, or something like it, is a crucial activity in the mechanism of spatial memory. The idea is essentially Hebb's (1949): when the presynaptic and the postsynaptic neuron are simultaneously active, the synapse is strengthened. LTP is commonly studied in the synapses of the mammalian hippocampus, a medial temporal lobe structure that is thought to be an important entity in the mechanism of spatial memory. A cross section of the hippocampus highlighting some of its major anatomical regions is shown in figure 6.2. Spatial memories are thought to be formed through the changing strengths of synapses between neurons in the hippocampus, and this is how LTP is thought to fit into the context of the mechanism of spatial memory.

The mechanisms of LTP are not yet completely understood. Robert Malinow, an LTP researcher, has worried in print that the LTC (long-term controversy) over LTP is becoming an LTTP (long-term tar pit) for neurobiologists (Malinow 1998, 1226). Nonetheless, one popular sketch of the mechanisms of LTP, visually represented in figure 6.3, includes the following organized collection of entities and activities. When the presynaptic neuron is active, it releases glutamate. This glutamate binds to N-methyl-D-aspartate (NMDA) receptors on the postsynaptic cell. The NMDA receptors change their conformation, exposing a pore in the cell membrane. If the postsynaptic cell is inactive, the channel remains blocked by large Mg^{2+} ions. But if the postsynaptic cell is depolarized, these Mg^{2+} ions float out of the channel, allowing Ca^{2+} to diffuse into the cell. The rising intracellular Ca^{2+} concentration sets in motion a long chain of biochemical activities terminating in the question marks of figure 6.3.

A number of gaps arise in the story at this point, but three things are thought to happen. In the short term, it is thought that this cascade leads to an increase in the number or sensitivity of α-amino-3-

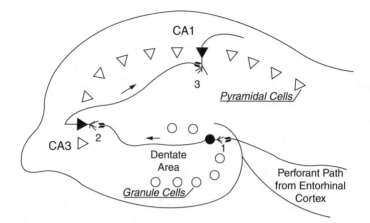

Figure 6.2 Cross section of the hippocampus. The CA1, CA3, and dentate regions are labeled. Numbers 1–3 denote the three synapses in the hippocampal trisynaptic loop. Circles and triangles represent granule cells and pyramidal cells, respectively.

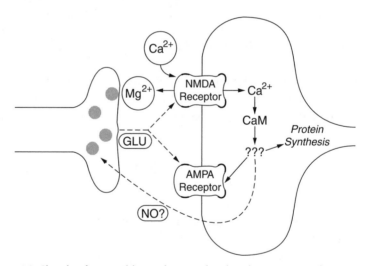

Figure 6.3 Sketch of a possible mechanism for the phenomenon of LTP.

hydroxyl-5-methyl-4-isoxazolepropionic acid (AMPA) receptors (perhaps by phosphorylation). These changes account for the rapid induction of LTP. In the long term, the cascade leads to the production of proteins in the postsynaptic cell body. These proteins are then used to alter the structure of the dendritic spines at that synapse. Some suspect that there is also a presynaptic component of the LTP mechanism, whereby, for example, the presynaptic cell releases more glutamate. Although incomplete, this description will suffice as a sketch of the mechanism of LTP. (More detail concerning this mechanism and the evidence that supports it can be found in Kuno 1995 and Frey and Morris 1998.)

The entities in this mechanism are glutamate molecules, NMDA receptors, Ca^{2+} ions, and the like. The activities include binding, diffusing, phosphorylating, and changing conformation. These entities and activities are organized in the production of LTP. The components exhibit a temporal organization that begins with the release of glutamate and terminates in structural changes that strengthen the synapse. The rates and durations of the different stages are crucial for the working of the mechanism; for example, there are the short-term modification of the AMPA receptors and the long-term structural changes to the dendritic spine. Stages of the mechanism are compartmentalized or localized in cells, membranes, and pores. Ignoring the question marks, the early stages of the mechanism are connected with one another, mostly through the motion, binding, and breaking of molecules. These molecular activities depend crucially on the structures and orientations of the entities involved; the size of the pore and the complementary shapes of glutamate and the NMDA receptor allow these entities to engage in the activities that produce the later stages of the mechanism.

Levels

There is often a third aspect to the organization of mechanisms in addition to the spatial and temporal; this is a hierarchical aspect.[1] Mechanisms in contemporary neurobiology are organized into multilevel hierarchies (figure 6.4). The mechanism of spatial memory is a good illustration. The description of this mechanism includes mice learning to navigate, hippocampi generating spatial maps, synapses inducing LTP, and macromolecules (like the NMDA receptor) binding and bending.

The levels in this sort of hierarchy stand in part-whole relations to one another with the important additional restriction that the lower-level entities and activities are components of the higher-level mechanism. The binding of glutamate to the NMDA receptor is a lower-level activity in the mechanism of LTP, and LTP is thought to be a lower-

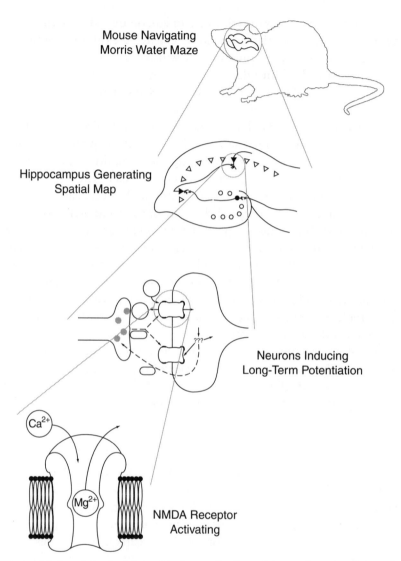

Mouse Navigating
Morris Water Maze

Hippocampus Generating
Spatial Map

Neurons Inducing
Long-Term Potentiation

NMDA Receptor
Activating

Figure 6.4 Levels in the hierarchical organization of the mechanism of spatial memory.

level activity in spatial map formation, which, of course, is thought to be an activity in the mechanism of spatial memory.

The elaboration and refinement of these hierarchical descriptions typically proceeds piecemeal with the goal of *integrating* the entities and activities at different levels. Integrating a component of a mechanism into such a hierarchy involves, first, contextualizing the item within the mechanism of the phenomenon to be explained. This involves "looking up" a level and finding a functional role for the item in that higher-level mechanism. There is some question, for example, over the role of LTP in the mechanism of spatial memory; not only is there a debate over the correct role for LTP in this mechanism, but there is even some debate as to whether it has a role at all. Contextualizing an item within a relevant mechanism for the phenomenon to be explained is one step in the integration of the different levels in a hierarchy.

The second means of integrating a component into such a hierarchical description is downward looking. This involves showing that the properties or activities of an entity can be explicated in terms of a lower-level mechanism. The persistent failure to find a mechanism for a postulated property or activity signals that there is an irremediable gap in the productive continuity of the mechanism. The activity or property cannot be integrated with those at lower levels. Thus integrating multilevel mechanisms involves both contextualizing an item within higher-level mechanisms and explicating that item in terms of lower-level entities and activities. (For more on the relationship between functions, mechanisms, and levels, see Craver 1998, 2001.)

Describing Mechanisms

The preceding discussion of mechanisms and their organization is motivated by the idea that thinking carefully about the abstract structure of mechanisms can provide insight into how they are discovered. Philosophical discussions of discovery should be sensitive to the fact that there are many different kinds of things to be discovered and that these different kinds of things are not all discovered in the same way. The product shapes the process of discovery.

The mechanism for a given phenomenon, for example, is typically not discovered all at once. Instead, descriptions of mechanisms are typically constructed piecemeal. Often neurobiologists understand some stages of the mechanism quite well and have only the sketchiest understanding of other stages. The question marks in the LTP diagram in figure 6.3 make these gaps in the mechanism explicit. The descriptions of the various components of a mechanism are often evaluated, and so revised, independently of others. In order to capture this feature of the discovery of mechanisms, it is useful to distinguish mechanism schemata from mechanism sketches.

Mechanism schemata are abstract descriptions of mechanisms that can be instantiated to yield descriptions of particular mechanisms. The term *mechanism schemata* is fitting because their components are placeholders that can be filled in with detailed stages between the setup and termination.[2] Schemata are thus complete in the sense that they can be filled in without gaps in the productive continuity of the mechanism. Schemata and their component placeholders typically have less than universal scope, and their scope can vary considerably.[3]

Mechanism sketches, in contrast to mechanism schemata, are abstract descriptions of mechanisms that cannot yet be filled in. Mechanism sketches have black boxes—they leave gaps in the productive continuity of the mechanism, such as the question marks in the LTP diagram. Such black boxes in mechanism sketches are useful in providing guidance about where further elaboration is needed. This role is especially important in the discovery of multilevel mechanisms.

The discovery of mechanisms unfolds gradually and piecemeal through the addition of constraints on plausible mechanism schemata and sketches. These constraints are used to construct plausible descriptions of mechanisms and to revise these plausible descriptions as constraints are added, deleted, or modified. Constraints determine the shape of the space of hypothesized mechanisms. Most simplistically, this space can be understood as a tree with terminal nodes representing possible mechanism schemata for the phenomenon to be explained. The addition of constraints prunes the tree or changes the weights on different branches. The removal of constraints, likewise, can add new branches to the hypothesis space. Understanding the discovery of mechanisms requires an understanding of these different constraints. This is the subject of the remainder of this chapter.

Constraints on the Organization of Mechanisms

Bechtel and Richardson's 1993 book *Discovering Complexity* discusses the use of localization and decomposition as research strategies in the discovery of mechanistic explanations and the conditions under which these strategies are prone to fail. Their book is an important contribution to the relatively sparse discovery literature in the philosophy of biology (see, e.g., Darden 1991; Darden and Cook 1994). Yet the contribution remains incomplete without a careful look at the products of this discovery process. Thinking carefully about mechanisms and especially their organization highlights a broad variety of constraints on their discovery in addition to those that come from localizing and decomposing. (Bechtel and Richardson's discussion of constraints on "causal and explanatory models" is in some ways complementary to our treatment; see 1993, 235.) For the remainder of the chapter, we focus on five varieties of constraint. These include the character of the phenomenon, componency constraints, spatial constraints, temporal constraints, and hierarchical constraints. Localization provides one kind of spatial constraint.

Characterizing the Phenomenon

Mechanism schemata and sketches are constrained by the character of the phenomenon for which a mechanism is sought. How one characterizes the phenomenon determines what will count as an adequate description of the mechanism that produces it; the complete description of the mechanism shows how that phenomenon is produced. Spatial memory is the phenomenon to be explained in our working example. But it is not at all obvious in advance of considerable empirical inquiry that there is any such individuable phenomenon—spatial memory—for which there exists an individuable mechanism; and given that there is such a faculty or phenomenon, it is not at all obvious in advance of considerable empirical inquiry how that phenomenon is properly to be characterized. Debates over the taxonomy of memory can be understood as debates about how to characterize and individuate different memory phenomena. The character of the phenomenon, like the description of the mechanism, is open to revision in light of evidence. This is a prevalent feature of the discovery of mechanisms.

Tolman's famous experiments on maze learning are a good example (Tolman and Honzick 1930; Tolman 1948). Tolman's work was instrumental in shaping the way in which contemporary neurobiologists think about spatial memory. Rats trained to navigate a circuitous route through a maze successfully were subsequently placed into the same maze with a new, more direct route from start to reward. If spatial memory were a simple association between stimulus and response, the rats would be expected to take the circuitous route for which they had been reinforced. But they did not; they preferred the more direct route. The rats could also construct efficient detours, shortcuts, and novel routes to the reward (see, e.g., Olton and Samuelson 1976; Chapuis et al. 1987). Importantly, these experiments and others like them suggest that spatial memory involves the formation of an internal spatial representation—a cognitive map—by which different locations and directions in the environment can be assessed. This characterization of the phenomenon guides the neurobiologist to seek out some entity, property, or activity in the central nervous system that could serve as a representation of space.

The characterization of the phenomenon is also shaped crucially by the accepted experimental protocols for producing, manipulating, and detecting it. As Bogen and Woodward (1988) have argued, phenomena should not be confused with data, which are the evidence for phenomena. Data, among other things, are idiosyncratic to particular experimental arrangements; phenomena, as we think of them, are the stable and repeatable properties or activities that can be detected, produced, and manipulated in a variety of experimental arrangements. For our present purposes, it is important to note that different experimental arrangements reveal different aspects of the phenomenon.

So it was mazes of differing complexity that led Tolman to think of spatial memory in terms of the formation of spatial maps. Spatial memory is also tested in radial arm mazes, three-table problems, and, most importantly for our purposes, the Morris water maze. The Morris water maze is a circular pool filled with an opaque liquid that covers a hidden platform. Mice are trained to escape over repeated trials. They do not like to swim, and so they learn quickly. The aquatic nature of the task also eliminates smell as a sensory cue. So the maze isolates the place of *visual information* in spatial memory. Sherry and Healy (1998, 133) underscore the importance of different experimental protocols for understanding the phenomenon for which a mechanism is sought: "The idea of the cognitive map, first proposed by Tolman (1948), has been an important and influential stimulus to research. But it is really more of a metaphor than a theory. Research on path integration, landmark use, the sun compass and snapshot orientation . . . attempts to specify more concretely exactly what makes up a 'cognitive map' of space." Different experimental assemblies accentuate different features of the phenomenon to be explained. Scientific debates often turn on the appropriateness of a given experimental arrangement for producing, manipulating, or detecting a given phenomenon. Debates over the ecological validity or ethological appropriateness of a task, for example, are debates over the character of the phenomenon. Experimental arrangements are often revised and adjusted over the course of the discovery of a mechanism.

Characterizing the higher-level phenomenon to be explained is a vital step in the discovery of mechanisms. Characterizing the phenomenon prunes the hypothesis space (since the mechanism must produce the phenomenon) and loosely guides its construction (since certain phenomena are suggestive of possible mechanisms). Yet such a top-down approach, as Bechtel and Richardson (1993, 237) agree, cannot itself exhaust the discovery of a mechanism. One also must know the components of the mechanism and how they are organized.

Componency Constraints

Mechanisms, remember, are composed of both entities and activities. For a given field at a given time there is typically a store of established or accepted components out of which mechanisms can be constructed and a set of components that have been excluded from the shelves. The store also contains accepted modules: organized composites of the established entities and activities. In contemporary neurobiology, for example, brain mechanisms will be composed of discrete neurons rather than a "reticulum." These neurons are connected by synapses, which may be electrical, chemical, or both. If they are chemical, then the mechanism will most likely involve action potentials, quantal release, and allosteric interactions. Modules in neurobiology include different second-messenger cascades, ionophore complexes, and cytoarchitectural structures, such as ocular dominance columns and glomeruli.

The store of entities, activities, and modules out of which mechanisms are constructed expands and contracts with the addition and removal of established entities and activities over time. Contracting the store adds constraints on plausible mechanisms by pruning those branches of the hypothesis space that represent mechanisms with such unestablished or unaccepted components. One commentator recently praised a hypothesized mechanism of LTP by saying, "If nothing else, this model is attractive because it requires only established intracellular signaling mechanisms" (Malinow 1998, 1226). Expanding the store of components loosens constraints by adding branches to the hypothesis space. The addition of nitric oxide (NO) to the store in the 1980s opened the hypothesis space to mechanisms involving retrograde transmission from the postsynaptic to the presynaptic neuron. It had previously been assumed that chemical neurotransmission was unidirectional, but this entity, which can diffuse freely through neuronal membranes, expanded the space of possibilities (figure 6.3). Importantly, the store of mechanism components provides guidance in the construction of mechanisms by supplying a set of ingredients out of which mechanisms might be concocted. Introductory neurobiology textbooks acquaint students with this store of entities, activities, and modules.

These textbooks also introduce students to various limitations on the activities in which these entities can engage. These features are important componency constraints on plausible mechanisms. For example, action potentials do not travel at the speed of light. The fastest move at 120 meters per second, and that is in the squid, with its appropriately named giant axons. Human action potentials propagate at roughly 1 meter per second, and our neurons can only fire 500 times in a second. Constraints can also be found in metabolic requirements, computational resources, temperature limits, rates of protein synthesis, and similar facts of carbon-based life on earth. A pair of authors recently dismissed the hypothesis that the proteins for altering the structure of the synapse in LTP were synthesized within each individual synaptic spine. They rejected this hypothesized mechanism as metabolically too demanding (Frey and Morris 1998, 182; they also discuss experimental evidence against the hypothesis). Although these componency constraints are always open to revision in the light of new evidence, they can be decisive in determining the fate of a proposed mechanism schema.

Spatial Constraints

Componency constraints shape the hypothesis space by delimiting the store of entities, activities, and modules that can be included in a mechanism and by limiting the possible activities in which the entities can engage. More specific constraints arise from empirical discoveries concerning the spatial organization of the mechanism. The components of mechanisms are often compartmentalized, localized, connected, structured, and oriented with respect to one another. Evidence concerning these sorts of spatial relationships among the components of a mechanism also constrains and guides the discovery process.

Often the components or stages of mechanisms are *compartmentalized* within reasonably well-defined regions. As the term suggests, these regions are often sectioned off by physical boundaries, like a nuclear membrane, a cell membrane, or skin. Compartmentalization often provides a natural way to individuate the stages of a mechanism. Transcription happens in the nucleus and translation happens in the cytoplasm; there are pre- and postsynaptic components of the mechanism of LTP. Compartmentalization also guides one to seek activities capable of linking the components inside this boundary with those outside—activities such as diffusion, active transport, and second messenger systems.

Closely related to compartmentalization is *localization*. Localizing components, the major focus of Bechtel and Richardson (1993), is often essential for understanding the spatial layout of a mechanism. For reasons to be discussed shortly, researchers are now reasonably confident that some components of the mechanism of spatial memory are to be found in the hippocampus. This finding opens up two new sets of research questions grounded in the spatial organization of the mechanism. One can, for instance, look inside the hippocampus to see what makes it work. Such investigation allows one to restrict the store of components to just those that can be identified within this spatial region. One is constrained to understand the activity of the hippocampus in terms of the cells, synapses, neurotransmitters, and circuits that can be found there.

One can then begin to describe the connections of these hippocampal components. *Connectivity* is yet a third variety of spatial constraint; the productive continuity of mechanisms relies upon the spatial connections among the components. Early research on the hippocampus has operated under the assumption that the anatomical connectivity of the hippocampal regions exhibits a characteristic clockwise "trisynaptic" loop (figure 6.2). Perforant path fibers from the entorhinal cortex make the first synapse onto granule cells in the dentate gyrus. These in turn project their axons to the pyramidal cells of the CA3 region, which in turn project to CA1 pyramidal cells. Revising this simple wiring diagram by adding new connections or new types of cells alters the space of plausible mechanisms by changing the scaffolding upon which the mechanism can be constructed. More recent research is beginning to emphasize the recurrent connections within these different regions.

Knowing that part of the spatial memory mechanism is localized to the hippocampus also constrains the description of the mechanistic context of the hippocampus. The anatomical connections into and out of the hippocampus become further constraints on the mechanism of spatial memory. This connected anatomy is the spatial scaffolding of the components of the mechanism. Localization is thus an important tool for revealing the connectivity of the mechanism.

Details concerning the geometrical *structure* and *orientation* of the entities in a mechanism, on the other hand, are important for understanding the productivity of mechanisms. As we noted earlier, the stages of mechanisms often depend crucially upon entities with appropriate structures having appropriate orientations with respect to one another. Discovering these structures and orientations can place important constraints on the hypothesis space. Indeed, structural aspects of the LTP mechanism are a major focus of recent research on LTP. Recent articles (Engert and Bonhoeffer 1999; Maletic-Savatic et al. 1999) present evidence for the addition of new dendritic spines to recently potentiated synapses. Although far from conclusive, such evidence argues for the existence of the structural basis for one plausible mechanism sketch for LTP.[4] The idea, yet to be confirmed, is that the addition of new dendritic spines makes the postsynaptic cell more responsive to glutamate. Structure thus provides clues to the activities that sustain the productive continuity of the mechanism. What remains to be shown, if this mechanism sketch is to be viable, is that the new postsynaptic spines are oriented properly with respect to the presynaptic axons.

The general point of this discussion is that mechanisms are organized spatially in the production of a phenomenon. Identifying aspects of that spatial organization guides and constrains the search through the hypothesis space in a number of ways that go beyond the strategy of localizing. Locations, boundaries, connections, positions, shapes, and orientations are especially important characteristics of, or relations among, the entities in mechanisms; these characteristics are especially important because they constrain the activities in which those entities can engage and so constrain the way that the mechanism can work. It is for this reason that characterizing these spatial aspects of the organization of mechanisms contributes to our understanding. But details of the spatial organization alone do not allow one to understand what mechanisms do. This spatial organization must be set in motion.

Temporal Constraints

Our efforts to understand just how a given mechanism moves are delimited and guided by knowledge of the mechanism's temporal organization. Knowledge of the order, rate, duration, and frequency of the activities in which the component entities of the mechanism engage provides clues to how the mechanism works.

Consider the temporal order of the activities of the entities composing a mechanism—that is, their relative position in the series, forks, and cycles that make up the mechanism. Spatial organization, in and of itself, does not reveal the direction of the productivity in the mechanism—the idea that the circuit of neurons in the hippocampus goes clockwise or that NMDA receptors allow Ca^{2+} influx into the postsynaptic cell, thereby initiating protein synthesis. Of course, temporal sequence is not by itself sufficient to establish these productive relationships, but given the temporal asymmetry of causality, temporal relations can place constraints on which entities and activities can be seen as productive of which others.

Temporal constraints on the discovery of mechanisms also include constraints imposed by the rates and durations of both the phenomenon to be explained and the stages of the mechanism. The speed limit for generating and propagating action potentials and for transmission at chemical synapses, for example, places limits on the number of sequential steps involved in a phenomenon of a given duration. Temporal constraints have been important in the discovery of the mechanisms of LTP. Researchers who believe that enduring LTP may be sustained by the addition of dendritic spines to the postsynaptic cell, for example, cannot use this mechanism to explain the initial induction of LTP because it takes around 30 minutes to produce the required proteins, distribute them, and insert them into the membrane. Short-term induction of LTP must rely on some faster mechanism, like the phosphorylation of AMPA receptors. Possible mechanisms are pruned from the hypothesis space on the grounds that the stages or steps take too long or happen too slowly to produce a phenomenon with a given rate or duration. Because mechanisms are active—because mechanisms do things—they take time to work, and the order, rate, and duration of the stages in a mechanism are therefore important tools for discovering their organization and culling the hypothesis space.

Experiments for Testing Hierarchical Mechanisms

The final set of constraints on the discovery of mechanisms is grounded in the hierarchical organization of mechanisms. Neurobiologists conduct experiments to reveal this hierarchical organization. Although there is a great deal to be said about how experiments are used to test mechanisms, the focus here is upon the role of experiments in the integration of levels in multilevel schemata and sketches.

Experiments can be understood in terms of an abstract experimental protocol, which clearly has some affinities with Hacking's (1988, 1992) discussion of experimentation.[5] Hacking does not discuss experimentation in the discovery of mechanisms per se, nor does he address their role in the integration of levels. We can achieve a more detailed understanding of experimentation by attending to the mechanistic organization that these experiments are used to reveal.

The abstract protocol in figure 6.5 is most easily described in the case of a unilevel mechanism; the protocol can then be easily extended to a multilevel case. The connected circles and arrows represent a hypothesized mechanism schema putatively instantiated in some experimental preparation. The schema may be instantiated in vivo, in vitro, or in silico (that is, in a mouse, in a petri dish, or in a computer). Having found or created such an experimental preparation, one then typically uses some *intervention technique* to perturb some component. The perturbation produced in the experimental preparation presumably has downstream results that are detected or amplified with the help of a *detection technique*. There is much to be said about this idealized protocol; we introduce it here simply to extend it to the multilevel case. And this is easy to do. Experiments designed to test the hierarchical organization of a mechanism typically involve intervening at one level and detecting at another. Sometimes, a single set of experiments involves intervening and detecting at multiple levels at once; we shall get to such a case shortly.

For simplicity, though, we start with two-level experiments. The left-hand side of figure 6.6 exhibits a case of intervening at the lower level and detecting at the higher level; we call these "bottom-up" experiments. The right-hand side of that figure shows the opposite:

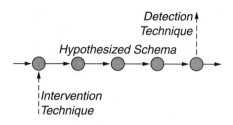

Figure 6.5 Abstract experimental protocol. Horizontal circles and arrows represent a hypothesized mechanism schema putatively instantiated in an experimental preparation. Vertical arrows represent techniques for intervening (left) to perturb the mechanism and for detecting (right) the results of that intervention.

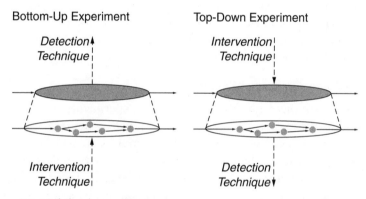

Figure 6.6 Multilevel experiments.

intervention at the higher level and detection at the lower level. These can be thought of as "top-down" experiments. Interventions, in either of these cases, may be either stimulatory or inhibitory. Our first examples are inhibitory, bottom-up experiments. Our second set of examples is top-down and stimulatory. Both sets of examples are drawn from experiments that, taken together, forcefully suggest that the hippocampus is involved in the formation of spatial maps.

Bottom-Up Inhibitory Experiments

The first example is the now-familiar case study H.M., as reported by Scoville and Millner (1957). Because H.M. was plagued by incapacitating epileptic seizures, he consented to an experimental surgical procedure to remove portions of his medial temporal lobes, including the hippocampus. After the surgery, it quickly became apparent that H.M.

had lost the ability to remember recent facts, even though he retained the ability to learn new skills. H.M.'s case famously suggested to researchers that the human hippocampus is a crucial entity in the mechanisms of what has since been called "declarative" memory (Zola-Morgan and Squire 1993).

Subsequent experiments in rats and mice have shown that bilateral removal of the hippocampus leads to profound deficits in spatial memory. For example, although rats with intact hippocampi learn very quickly, over repeated trials, to swim directly to the hidden platform in the Morris water maze, rats with bilateral hippocampal lesions continue over multiple trials to swim randomly through the pool, stopping only when they stumble onto the platform (Morris et al. 1982).

Both the case of H.M. and the subsequent ablation experiments in mice are examples of the two-tiered experimental structure exhibited in the left-hand side of this figure.

Top-Down Excitatory Experiments

The findings of these inhibitory, bottom-up experiments are reinforced by excitatory, top-down experiments like those on the right-hand side of the figure. In the early 1970s, O'Keefe and Dostrovsky (1971) recorded the electrical potentials of individual pyramidal cells in the CA1 region of the rat hippocampus while rats navigated a standard maze. The intervention in this case involves activating the spatial memory system by putting the rat in a maze. The detection technique is the electrical recording. They found that certain of those pyramidal cells generate bursts of action potentials whenever the rat enters a particular location while facing in a particular direction. These cells have come to be called "place cells," and the region of space occupied by the rat when the place cell increases its activity is likewise known as the cell's "place field." The place cells of CA1 have slightly overlapping place fields that cover the animal's immediate spatial environment. The pattern of activity across this subpopulation of CA1 pyramidal cells could therefore play the role of a spatial map.

These findings have recently been confirmed with multiunit electrodes that allow one to record from 70 to 150 CA1 pyramidal cells at once. Astonishingly, it is possible to *predict* the path taken by the rat on the basis of these recordings (Wilson and McNaughton 1993). This is a remarkable top-down stimulatory finding.

Top-down and bottom-up experiments of both the stimulatory and inhibitory variety are quite common in neurobiology. They are common because the findings of such experiments, taken together, reveal aspects of, and thereby place constraints upon, the hierarchical organization of a mechanism. More specifically, when experiments of this sort go well, they place constraints on the possibilities for integrating the different levels. Top-down and bottom-up experiments help to situate an item, like the hippocampus or the NMDA receptor, within the context of a higher-level mechanism. They also identify components in the mechanisms that produce higher-level activities and properties. These experiments tell us what the relevant entities and activities are, how they are nested in component-subcomponent relations, and how the activities of the component entities fit into their mechanistic context. Persistent failure to situate an item within a hierarchical mechanism, or persistent failure to uncover a lower-level mechanism for that item, prunes mechanism schemata involving that item from the hypothesis space.

This role of these experiments in placing constraints on the integration of the different levels of a hierarchy is even more apparent in multilevel experiments, from which our last example is drawn.

Multilevel Experiments

In late 1996, researchers at MIT, Columbia, and Cal Tech published a series of papers describing the effects of highly specific genetic deletions (or "gene knockouts") on entities and activities at multiple levels in the spatial memory hierarchy (McHugh et al. 1996; Rottenberg et al. 1996; Tsien et al. 1996a, 1996b).

The experiments that are our focus are bottom-up and inhibitory. Specifically the researchers invented a molecular scalpel for deleting the *NMDAR1* gene, a gene encoding an essential subunit of the NMDA receptor (Tsien et al. 1996a), and for deleting it only in the pyramidal cells of the CA1 region of the mouse hippocampus. The deletion was also timed to occur only after normal hippocampal development is thought to be complete. The trick was to couple the deletion to a promoter of a gene that is expressed selectively in CA1 pyramidal cells and that is expressed only in the later developmental stages of the hippocampus. This intervention technique gives researchers finer-grained spatial and temporal resolution in their manipulation of the brain's activities than has ever been possible. This, in turn, provides higher spatio-temporal resolution on the organization of the mechanism of spatial memory.

These experiments have been praised as the first to investigate the mechanism of spatial memory, "at all levels in a single set of experiments, from molecular changes through altered patterns of neuronal firing to impaired learning" (Roush 1997, 32) and for taking an important first step toward the "dream of neurobiology . . . to understand all aspects of interesting and important cognitive phenomena—like memory—from the underlying molecular mechanisms through behavior" (Stevens 1996, 1147). More specifically, we claim that these experiments advance the goal of integrating the different levels in this multilevel mechanism.

Knockout mice, those without functional NMDA receptors, had difficulty escaping the Morris water maze. They performed far worse than controls in learning to escape. When placed in a maze *without* a platform, control mice concentrated their swimming in the platform's previous location. Knockout mice swam about randomly (Tsien et al. 1996b). Multiunit recordings from CA1 pyramidal cells in the knockout mice revealed significant impairments in spatial map formation. The researchers found, to their surprise, that CA1 cells in the knockout mice *did* exhibit place-related firing. But the place fields were much larger and much less sharply defined. These deficits in spatial map formation can reasonably be attributed to the absence of LTP at synapses lacking functional NMDA receptors. The researchers found that knocking out the NMDA receptor eliminated LTP induction in CA1 and not in any other region of the brain (Tsien et al. 1996b).

This complicated experiment is a bottom-up inhibitory experiment with detection at multiple levels. The intervention technique intervenes to perturb the NMDA receptor by deleting the *NMDAR1* gene. The detection techniques register the results of this intervention on LTP, spatial map formation, and spatial memory. There is a lot to be said about the strength of these experimental findings, but this is not our focus here. Instead we are interested in how these multilevel experiments constrain hypotheses about the integration of multilevel mechanisms.

This set of experiments is designed to test a popular sketch of the multilevel mechanism of spatial memory. It is a sketch because we are not remotely in a position to trace out all of the mechanisms at all of the different levels. Instead, this sketch is a hypothesis of how the components at different levels are integrated with one another. In particular, it is the hypothesis that the NMDA receptor is a necessary component of the mechanism of LTP, which is a necessary component of the mechanism of spatial map formation, which is a necessary component of the mechanism of spatial memory. If these nesting relationships do hold, then knocking out an essential gene for the NMDA receptor would be expected to eliminate the induction of LTP, to eliminate spatial map formation, and to leave the mice hopelessly lost in the Morris water maze. The rough conformity of the findings to these expectations is heartening in this respect.

It is important to note, however, that the genetic deletion did not eliminate spatial map formation in the CA1 region of the hippocampus. Instead, knocking out this essential gene made the map less precise. This lack of precision still had behavioral implications, and so it is consistent with some role for LTP within the context of the spatial memory mechanism as a whole. However, this anomalous finding forced the researchers to rethink the role of CA1 LTP in the context of this mechanism. The persistence of place-related firing in CA1 suggests that place fields must be established in an earlier stage of the mechanism. So the role of CA1 within the context of the spatial memory mechanism is not the formation of spatial maps. Instead the researchers suggest, with characteristic caution, that CA1 has the role of "learn[ing] associations between entorhinal inputs and place [information projecting from the CA3 region of the hippocampus]" (McHugh et al. 1996, 1347). This finding, in other words, constrains our understanding of the role of CA1 LTP in the spatial memory mechanism; as this case of revision suggests, it also helps to establish the productive organization of the components of the mechanism as a whole. (For a more systematic discussion of anomaly resolution, see Darden 1991, 1992; Darden and Cook 1994.)

It is in this way that multilevel experiments furnish constraints on the integration of components at multiple levels. These hierarchical constraints, in conjunction with the character of the phenomenon and the spatial, temporal, and componency constraints, guide and delimit the discovery of mechanisms.

Conclusion

It is now possible to step back and look at this discovery process as a whole. The continuing discovery of the mechanism of spatial memory has proceeded gradually through the piecemeal elaboration and revi-

TABLE 6.1 Summary of Constraints on
the Discovery of Mechanisms Exhibited
in the Discovery of the Mechanism of
Spatial Memory

1. Character of the phenomenon
2. Componency of constraints
Store of entities and activities
Modules
3. Spatial constraints
Compartmentalization
Localization
Connectivity
Structure
Orientation
4. Temporal constraints
Order
Rate
Duration
Frequency
5. Hierarchical constraints
Integration of levels

sion of mechanisms at multiple levels and through the gradual integration of the components at each of these levels. This discovery process has been guided by a sketch, replete with black boxes, of how this mechanism is organized. Certain of these black boxes are starting to open with the accumulation of constraints on how they are to be filled in. We are beginning to understand the details of LTP, and perhaps more importantly, we are beginning to evaluate more precisely the role of LTP within the context of the spatial memory mechanism. The same is true for spatial map formation in the hippocampus and the opening and closing of the NMDA receptor.

This discovery process is guided by constraints on the organization of the mechanism that are summarized in table 6.1. These constraints come from many different specialties within neurobiology: neuroanatomists, clinical psychologists, electrophysiologists, and molecular neurobiologists are all contributing different constraints on the emerging organization of this hierarchical mechanistic structure. Discussions of scientific discovery in neurobiology should proceed by attending to the organizational structure of mechanisms. The product shapes the process of discovery. In understanding both the product and the process we come to see more clearly what is involved in understanding phenomena by describing mechanisms.

NOTES

This chapter is based on work supported by the National Science Foundation under grant SBR-9817942. Any opinions, findings, and conclusions or recommendations are those of the authors and do not necessarily reflect the views of the National Science Foundation. Craver's work was also partly funded by the Committee on Cognitive Studies Postdoctoral Fellowship in the Department of Philosophy at the University of Maryland, College Park. We thank German Barrionuevo, Nancy Hall, Tetsuji Iseda, Peter Machamer, Gualtiero Piccinini, Rob Skipper, and Nathan Urban for useful comments on earlier drafts of this chapter and Payman Farsaii for help as an undergraduate research assistant.

1. Although the notion of "hierarchy" is often associated with the control or governance of things at lower levels by those at higher levels, this should in no way be associated with the notions of "level" and "hierarchy" explicated here.

2. Skipper (1999) develops a mechanism schema for selection mechanisms.

3. Schaffner (1993) has suggested that the "bulk" of theories in the biomedical sciences be seen as "overlapping interlevel temporal models of varying scope." Although we are sympathetic with the direction of Schaffner's thinking here, we prefer to think of such theories as schematic multilevel descriptions of mechanisms. Our discussion of spatial, temporal, and hierarchical constraints on descriptions of mechanisms is intended to exhibit the additional content of an explicit emphasis on mechanisms over and above less specific talk of "theories" or "models." Both mechanism schemata and their components (the placeholders for entities and activities) can have widely varying scope, from near-terrestrial universality (such as the mechanism of protein synthesis) to mechanisms, entities, or activities that are found only in some parts of some highly specific strains of organisms. (More on scope can be found in Darden's [1996] review of Schaffner [1993].) Because the sense of "level" articulated in the second section is explicitly defined in terms of "componency" and hence "mechanism," emphasis on mechanisms in the structure of these theories also brings with it a sensible interpretation of their multilevel character. For discussions of multilevel mechanism schemata as theories in neurobiology, see Craver (1998, 10–48; 2001).

4. Additional details concerning this hypothesized mechanism continue to emerge. See Barinaga (1999) and Shi et al. (1999) on the delivery and activation of AMPA receptors in LTP.

5. For an alternative schematic account of experiments, see Lederberg (1995).

REFERENCES

Barinaga, M. 1999. "New Clues to How Neurons Strengthen Their Connections." *Science* 284: 1755–57.

Bechtel, W., and R. C. Richardson. 1993. *Discovering Complexity: Decomposition and Localization as Strategies in Scientific Research*. Princeton, N.J.: Princeton University Press.

Bogen, J., and J. Woodward. 1998. "Saving the Phenomena." *Philosophical Review* 97: 303–52.

Chapuis, N., M. Durup, and C. Thinus-Blanc. 1987. "The Role of Exploratory Experience in a Shortcut in Golden Hamsters (*Mesocricetus auratus*)." *Animal Learning and Behavior* 15: 174–78.

Craver, C. F. 1998. "Neural Mechanisms: On the Structure, Function, and Development of Theories in Neurobiology." Doctoral dissertation, Department of History and Philosophy of Science, University of Pittsburgh, Pittsburgh, Pennsylvania.

———. 2001. "Role Functions, Mechanisms, and Hierarchy." *Philosophy of Science*, March.

Darden, L. 1991. *Theory Change in Science: Strategies from Mendelian Genetics*. New York: Oxford University Press.

———. 1992. "Strategies for Anomaly Resolution." In R. Giere, ed., *Cognitive Models of Science*. Minnesota Studies in the Philosophy of Science, vol. 15. Minneapolis: University of Minnesota Press, 251–73.

———. 1996. "Generalizations in Biology." *Studies in History and Philosophy of Science* 28: 409–19.

Darden, L., and M. Cook. 1994. "Reasoning Strategies in Molecular Biology: Abstractions, Scans, and Anomalies." In D. Hull, M. Forbes, and R. M. Burian, eds., *PSA 1994*, vol. 2. East Lansing, Mich.: Philosophy of Science Association, 179–91.

Engert, F., and T. Bonhoeffer. 1999. "Dendritic Spine Changes Associated with Hippocampal Long-Term Synaptic Plasticity." *Nature* 399: 66–70.

Frey, U., and R. G. Morris. 1998. "Synaptic Tagging: Implications for Late Maintenance of Hippocampal Long-Term Potentiation." *Trends in Neuroscience* 21: 181–88.

Hacking, I. 1988. "On the Stability of the Laboratory Sciences." *Journal of Philosophy* 85: 507–14.

———. 1992. "The Self-Vindication of the Laboratory Sciences." In A. Pickering, ed., *Science as Practice and Culture*. Chicago: University of Chicago Press, 29–64.

Hebb, D. O. 1949. *The Organization of Behavior*. New York: Wiley.

Kuno, M. 1995. *The Synapse: Function, Plasticity, and Neurotrophism*. Oxford: Oxford University Press.

Lederberg, J. S. 1995. "Notes on Systematic Hypothesis Generation and Application to Disciplined Brainstorming." In *Working Notes: Symposium: Systematic Methods of Scientific Discovery*. AAAI Spring Symposium Series. American Association for Artificial Intelligence, Stanford University, 97–98.

Machamer, P., L. Darden, and C. F. Craver. 2000. "Thinking about Mechanisms." *Philosophy of Science* 67: 1–25.

Maletic-Savatic, M., R. Malinow, and K. Svoboda. 1999. "Rapid Dendritic Morphogenesis in CA1 Hippocampal Dendrites Induced by Synaptic Activity." *Science* 283: 1923–26.

Malinow, R. 1998. "Silencing the Controversy in LTP?" *Neuron* 21: 1226–27.

McHugh, T. J., K. I. Blum, J. Z. Tsien, S. Tonegawa, and M. A. Wilson. 1996. "Impaired Hippocampal Representation of Space in CA1-Specific NMDAR1 Knockout Mice." *Cell* 87: 1339–49.

Morris, R. G. M., P. Garrud, J. N. P. Rawlins, and J. O'Keefe. 1982. "Place Navigation Impaired in Rats with Hippocampal Lesions." *Nature* 297: 681–83.

O'Keefe, J., and J. Dostrovsky. 1971. "The Hippocampus as a Spatial Map: Preliminary Evidence from Unit Activity in the Freely Moving Rat." *Brain Research* 34: 171–75.

Olton, D. S., and R. J. Samuelson. 1976. "Remembrances of Places Passed: Spatial Memory in Rats." *Journal of Experimental Psychology: Animal Behavior Processes* 2: 97–116.

Rottenberg, A., M. Mayford, R. D. Hawkins, E. R. Kandel, and R. U. Muller. 1996. "Mice Expressing Activated CaMKII Lack Low Frequency LTP and Do Not Form Stable Place Cells in the CA1 Region of the Hippocampus." *Cell* 87: 1351–61.

Roush, W. 1997. "New Knockout Mice Point to Molecular Basis of Memory." *Science* 275: 32–33.

Schaffner, K. 1993. *Discovery and Explanation in Biology and Medicine.* Chicago: University of Chicago Press.

Scoville, W. B., and B. Millner. 1957. "Loss of Recent Memory after Bilateral Hippocampal Lesions." *Journal of Neurology, Neurosurgery, and Psychiatry.* 20: 11–20.

Sherry, D., and S. Healy, 1998. "Neural Mechanisms of Spatial Representation." In S. Healy, ed., *Spatial Representation in Animals.* Oxford: Oxford University Press, 133–57.

Shi, S., Y. Hayashi, R. S. Petralia, S. H. Zaman, R. J. Wenthold, K. Svoboda, and R. Malinow. 1999. "Rapid Spine Delivery and Redistribution of AMPA Receptors after Synaptic NMDA Receptor Activation." *Science* 284: 1811–16.

Skipper, R. 1999. "Selection and the Extent of Explanatory Unification." *Philosophy of Science* 66: S196–S209.

Stevens, C. F. 1996. "Spatial Learning and Memory: The Beginning of a Dream." *Cell* 87: 1147–48.

Tolman, E. 1948. "Cognitive Maps in Rats and Men." *Psychological Review* 55: 189–208.

Tolman, E., and C. Honzick. 1930. "Introduction and Removal of Reward and Maze Performance in Rats." *University of California Publications in Psychology* 4: 257–75.

Tsien, J. Z., D. F. Chen, D. Gerber, C. Tom, E. Mercer, D. Anderson, M. Mayford, and E. R. Kandel. 1996a. "Subregion- and Cell Type–Restricted Gene Knockout in Mouse Brain." *Cell* 87: 1317–26.

Tsien, J. Z., P. T. Huerta, and S. Tonegawa. 1996b. "The Essential Role of Hippocampal CA1 NMDA Receptor–Dependent Synaptic Plasticity in Spatial Memory." *Cell* 87: 1327–38.

Wilson, M. A., and B. McNaughton. 1993. "Dynamics of the Hippocampal Ensemble Code for Space." *Science* 261: 1055–58.

Zola-Morgan, S., and L. Squire. 1993. "Neuroanatomy of Memory." *Annual Review of Neuroscience* 16: 547–63.

7

The Explanatory Power and Limits of Simulation Models in the Neurosciences

Holk Cruse

Faculty of Biology, University of Bielefeld, Bielefeld, Germany

The ultimate goal of neurobiological investigation is to understand how the brain contributes to the control of behavior. Brains of humans and higher mammals are considered to be the most complex systems in the world. Therefore reaching this goal is surely not a simple task. Even so-called "simple" brains, such as those of insects, are far from being understood. Important progress has been made with respect to specific, typically low-level questions by following a reductionistic approach. However, significant progress in understanding the whole system can be expected only if these reductionist methods are complemented by the "antireductionistic" or synthetic tools of simulation. In simulation studies, the separately investigated elements of the system can be put together in order to examine system properties that emerge from the interaction of these elements. These synthetic studies are normally performed on the basis of computer simulations. However, many properties of the system, including those of the environment in which the system is acting, cannot easily be simulated. The environment is very important, because, as already mentioned, the brain only *contributes* to the control of the behavior, yet does not completely *determine* behavior. Behavior results from the cooperation between brain and environment. Therefore, in recent years hardware simulations—in which the simulation takes the form of a real, physical robot often referred to as an animat (Wilson 1991; Dean 1998)—have assumed an important role. In this chapter I cite examples from our

138

research on insect walking, illustrating how we use both software and hardware simulations that are based on extensive experimental investigation of biological systems.

Computer Simulations

A computer simulation represents a quantitative model of the real world. It is helpful for at least three reasons: First, a model is able to provide a concise description of a large quantity of data. A mean value, a single number that represents many data points, is an example of such a model. A more complicated and therefore more interesting example is provided in the next section. This example also illustrates the second reason why quantitative models are useful: such models can represent a hypothesis concerning the mechanisms that may underlie the observed phenomenon. In this way the model provides not only a description but also a possible explanation. Third, such a quantitative hypothesis has predictive value in that it can lead to new experiments, the results of which may or may not support the hypothesis (this will be discussed in a subsequent section).

Leg Coordination in Crayfish

I illustrate these three properties by means of an example, which deals with the control of walking in a multilegged animal such as an insect or a crayfish. A system controlling many legs must solve two basic problems. The first is the control of the quasi-rhythmic movement of the individual leg, switching between the *stance phase,* in which the leg is on the ground supporting the body, and the *swing phase,* in which it is lifted off the ground and moves in the direction of walking to the point at which it can begin a new stance. The second basic problem concerns the coordination of multiple legs. Animals walk in clearly distinguishable gaits, which are stable even when external disturbances appear. How can a system be constructed that produces a stable spatio-temporal pattern and, at the same time, tolerates disturbances? To investigate the properties of such a system, we observed the reactions of real biological systems to different disturbances, for example, brief interruptions of the stance movement of one leg. Figure 7.1 shows that, in the crayfish, the legs return to normal coordination by shortening or prolonging the swing or the stance movements, or both, of some of the neighboring legs.

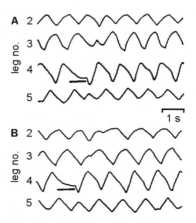

Figure 7.1 The coordinated walk of four ipsilateral legs of a crayfish is disturbed by interruption of the stance movement of leg 4 (bar). Following this disturbance, the system regains normal coordination by changing the step amplitude of the other legs.

The collection of a large amount of data from real crayfish led to the hypothesis that two coordinating mechanisms are active. One is rostrally directed. It prolongs the swing of a leg as long as its posterior neighbor performs a stance. The other is caudally directed. When the anterior leg is close to the end of its stance or at the beginning of its swing phase, this second influence has the effect of ending the swing of the posterior leg and beginning the stance of this leg. (A third influence, acting between each contralateral pair of legs, will not be considered here.)

Computer Simulations of Crayfish Coordination

Are these two mechanisms sufficient to explain the coordination patterns found in walking crayfish? Four ipsilateral legs connected by six individual influences form a system with a recurrent flow of information (figure 7.2). Experience shows that humans have difficulty imagining the dynamic behavior of systems with recurrent connections. However, the properties of such a hypothetical system can easily be studied by means of a simulation. In our case the results have shown that with this assumption it is possible to describe the behavior of a crayfish even in conditions marked by disturbances. This means, first, that the com-

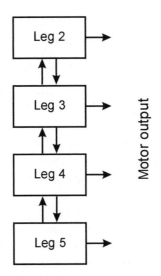

Figure 7.2 Four leg controllers (boxes) are connected by two different coordinating mechanisms (arrows), forming a system with recurrent connections.

bination of these two simple local rules is able to describe a large amount of data, and, second, that these rules form a sensible hypothesis concerning the mechanisms underlying the coordination of leg movement.

Furthermore, these rules describe not only the data that have been used to formulate the hypothesis. For example, in a normally walking animal we eventually observed very small steps intercalated between the normal steps. Our immediate intuition was to apply an additional mechanism to describe this behavior. However, the existing simulation has revealed very similar intercalated steps, which occur in the same particular leg configurations as they do in the animals. Thus the simulation shows some predictive value.

The fact that a small number of local rules is able to produce a stable and tolerant global behavior shows that this decentralized control system is an interesting solution, but it does not prove that this really is the solution applied by the animals. Nor does it show how these rules may be realized on the lower (neuronal) level. The model basically provides proof of feasibility and, in our case, shows some additional plausibility because more results were described than had been used to formulate the hypothesis in the first place.

Control of Leg Movement in Stick Insects

Let us now turn to another animal, the stick insect. In this animal we have found not two, but six different types of coordinating mechanisms on the basis of observation of real biological systems. Therefore a simulation to test the properties of the complete system is even more important, and one has in fact been performed successfully. I focus here on the problem of controlling leg movement during the stance phase (for details see Cruse et al. 1998a).

To show that control of walking is by no means an easy task, consider the problems to be solved. During stance it is not sufficient to specify a movement for each leg on its own, as in the case of the swing movement: the movement of the legs in the stance phase is constrained because they are mechanically coupled via the body and via the substrate. Therefore efficient locomotion requires coordinated movement of all the joints of all the legs in contact with the substrate, that is, a total of eighteen joints when all the legs of the insect are in contact with the ground. The task is a nonlinear problem, because we must deal with rotational movements and because the number and combination of mechanically coupled joints vary from one moment to the next, depending on which legs are lifted.

For straight walks, one could simplify the problem by assuming that the trajectories of the leg endpoints follow a straight line parallel to the long axis of the body. This assumption, however, is only approximately valid for normal walking. It definitely does not hold when the animal negotiates a curve, which requires the different legs to move along different arcs at different speeds. Superimposed on these problems is the fact that the control system must deal with extra degrees of freedom: that is, it must decide, by applying some criterion or criteria, which of the possible solutions should be selected.

In machines these problems can be solved using traditional (though computationally costly) methods, which consider the ground reaction forces of all legs in stance and seek to optimize some additional criterion or criteria, such as minimizing the tension or compression exerted by the legs on the substrate. Due to the nature of the mechanical interactions (and inherent in the search for a globally optimal control strategy), such algorithms require a single, central controller; they do not lend themselves to distributed processing. This makes real-time control difficult, even in the simple case of walking on a rigid substrate. Taking into account the much smaller bandwidth and the much slower computational speed of biological systems compared with electronic ones, it is evident that real-time control via such a central mechanism would be even more difficult.

Further complexities arise in more irregular, natural walking situations, making a solution difficult even for systems with significant computational power. These problems arise, for example, when an animal or a machine walks on a slippery surface or on a compliant substrate, such as the leaves and twigs encountered by stick insects. Any flexibility in the suspension of the joints further increases the degrees of freedom that must be considered and the complexity of the computation. Further problems standing in the way of an exact, analytical solution arise when the length of leg segments changes during growth or their shape changes through injury. In such cases, knowledge of the geometric situation is incomplete, making an explicit calculation difficult, if not impossible. Such problems even arise during normal walking: the positions and orientations of the axes in the nonrigid joints may change owing to load changes caused by the assumption of different orientations with respect to gravity.

Despite the evident complexity of these tasks, they are mastered even by insects, with their "simple" nervous systems. Hence there must be a solution that is fast enough that online computation is possible even for slow neuronal systems. How can this be achieved? Several authors (e.g., Brooks 1995) have pointed out that some relevant parameters do not need to be explicitly calculated by the nervous system because they are already available in the interaction with the environment. This means that, instead of an abstract calculation, the system can directly exploit the dynamics of the interaction and thereby avoid a slow, computationally exact algorithm. To solve the particular problem at hand, we propose to replace a central controller with distributed control in the form of local positive feedback (Cruse et al. 1998a). Compared with earlier versions of the control system (Müller-Wilm et al. 1992), this change permits that part of the net that is responsible for the control of stance movement to be radically simplified. The positive feedback occurs at the level of single joints: the position signal of each is fed back to control the motor output of the same joint. How does this system work? Let us assume that any one joint is moved actively. Then, because of the mechanical connections, all other joints begin to move passively, but in exactly the proper way. Thus the direction of movement and speed of each joint need not be computed because this information is already provided by the physics of the system. The positive feedback then transforms this passive movement into an active movement.

Although this argument may appear quite sensible, it is difficult to imagine how the complete system really behaves. The only realistic possibility for testing the feasibility of this idea is to perform a simulation. As the proposed control principle relies on the existence of a body, the mechanics of the body must also be simulated. To simulate the body, a recurrent network called MMC (for mean of multiple computations) has been used (see Steinkühler and Cruse 1998 for details).

To test this hypothesis, a network called Walknet has been developed to control the body. This network is based on the above-mentioned principle of local positive feedback, but it includes some additional elements not discussed in detail here, for example, the coordinating mechanisms also mentioned. As is shown in figure 7.3A for the case of straight walking, this network is able to achieve proper coordination. Steps of ipsilateral legs are organized in triplets forming "metachronal waves," which proceed from back to front, whereas the contralateral legs on each segment step approximately in alternation. With increasing walking speed, a change in coordination pattern from a typical tetrapod- to a tripod-like gait is found. For slow and medium velocities, the walking pattern corresponds to the tetrapod gait, with four or more legs on the ground at any time and diagonal pairs of legs stepping approximately together; for higher velocities, the gait approaches the tripod pattern, with front and rear legs on each side stepping together with the contralateral middle leg. The coordination pattern is very stable. For example, when the movement of one leg is interrupted briefly during the power stroke, normal coordination is regained immediately at the end of the perturbation. Furthermore, the model can cope with obstacles higher than the normal distance between the body and the substrate. It continues walking when a leg has been injured, for example, when half of a tibia is removed. It also negotiates curves very much like real insects do (as shown in figure 7.3B).

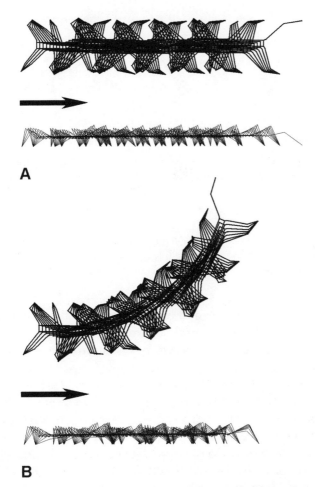

Figure 7.3 Simulated walking by the basic six-legged system with negative feedback applied to the six joints that are responsible for the control of body height and positive feedback to all other joints. The direction of movement is from left to right (arrow). Leg positions are illustrated only during stance and only for every second time interval in the simulation. Each leg makes about five steps. Upper trace: top view; lower trace: side view. (A) Straight walking; (B) curved walking.

Unexpectedly, the following interesting behavior was observed. A massive perturbation, for example by clamping the tarsi of three legs to the ground, can make the system fall (figure 7.4). Although this can lead to extremely disordered arrangements of the six legs, the system

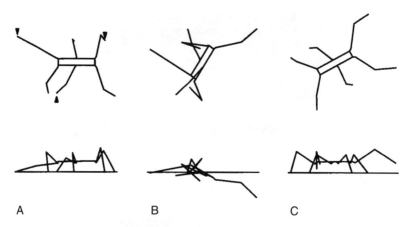

A   B   C

Figure 7.4 Righting behavior. (A) By clamping the tarsi to the ground (arrowheads), the system is made to fall, leading to a disordered arrangement of the legs (B). Nevertheless, the system stands up without help and resumes proper walking (C).

was always able to stand up and resume proper walking without any help. This means that the simple solution proposed here also eliminates the need for a special supervisory system to rearrange leg positions after such an emergency.

Considering all the problems a walking system must already deal with under seminatural conditions, one is inclined to attribute rather high motor intelligence to such a system.[1] We have, however, seen that the control system need not be very complicated, nor does it require a centralized architecture. On the contrary, the seemingly most difficult problems are solved by the structurally simplest subsystems. This simplification is possible because the physical properties of the system and its interaction with the world are exploited to replace an abstract, explicit computation. No explicit internal world model is required. Thus "the world is its own best model" (Brooks 1995, 54).

Further Research

Although the simulation just described has proved successful, it does not demonstrate that the proposed mechanisms are actually able to control a real physical system. The major disadvantage of our simulation is its purely kinematic nature; forces like inertia or friction are neglected. It is very difficult to perform a realistic "dynamic" simulation that includes all these physical properties.[2] It is even more difficult to simulate a realistically complex (i.e., natural) environment. One of the problems is that one cannot be sure if any important properties have been missed. There is, however, a theoretically simple solution to this kind of problem. This is to produce a simulation that does not exist in a "pixel world" (as do the computer simulations) but in the real, physical world—a "hardware simulation," that is, a robot or an animat. When a physically realized animat acts in the real world, all arguments concerning possible flaws or omissions in the underlying software simulation are rendered irrelevant. In our case, the hypothesis of local positive feedback will be investigated in cooperation with Martin Frik of the University of Duisburg, using his robot TARRY.

After we have determined whether the application of local positive feedback can be verified, what will be the next step in this research project? At present we are investigating the question of whether the behavior of insects is controlled by motivations. Do insects have internal states that affect their behavior such that the way the insect reacts to a given stimulus depends on its internal state? Preliminary results (Cruse et al. 1998a) support this assumption. After having finished these experiments we are again attempting to introduce motivational structures, first into the software simulations and later into the animat.

The Scope of Simulation Studies

Issues Concerning the Level of Simulation

In the simulations described in the previous sections, we decided to study the question of leg coordination on the behavioral level, but we might as well have studied it on the neuronal level or even on the molecular level. It could always be argued that one or the other approach is better suited to describe "reality." I challenge this view by noting that all these levels of description are different insofar as each of them illustrates some properties of the system better than others. Because of our limited perceptual abilities we cannot experience the complete reality of a given system, only selected aspects of it. For example, we would not have been very likely to discover the two local rules described above had we studied the system only on the molecular level. On the other hand, as previously stated, no information concerning the cellular or the molecular level is available from the behavioral studies. However, studying a system on one level may help to understand it on another level, given that both levels are not completely distinct. This view is of general interest for the field of neuroscience, with its studies on the molecular level, studies on the cellular level, studies of small cell assemblies, studies of complete brains, studies of "simple" behavior, on up to studies on the psychological level and even of social structures. None of these levels is more "real" than any other.

This situation may be illustrated by the following simple example. Imagine a computer program written to display some geometrical figure on the screen, for example, a rotating colored cube. Such a program may be written in different forms (figure 7.5). In an object-oriented language there would be easily understandable expressions such as "rotate()" or "shadow()." The program might, however, also be written in a classical language, such as Pascal or C, or in assembly language. And finally everyone knows that the "real" code is a binary one, consisting of a long series of the numbers zero and one. All these different versions describe the same procedure. Which of them describes what is really going on when the program is executed? This question is not sensible, because what is really going on is a physical process. All of the different versions of the program describe the same procedure, but they illustrate different properties of it. It is not possible, even for an experienced programmer, to understand what the program is doing if he or she can study it only on the binary level. Yet it is easy to do so when it is viewed at a higher level. However, when the goal is to understand the underlying lower-level processes, a review of the assembly code may be more appropriate. But no version of the program is "truer" than any other.

The Interaction of Simulation and Experiment

As mentioned in the section on "Computer Simulations," one benefit of computer simulations is that they produce testable hypotheses. As an example, consider the problem of the control of leg movement during the swing phase. After having studied the swing movement of a walking insect, we performed a simulation that described the three-dimensional movement of a leg. An extremely simple artificial neural network was developed that controlled the velocity of all three leg joints. This network successfully described normal swing movements and the behavior of the leg when it was mechanically disturbed during swing. For very small step amplitudes, which seldom occur in normal

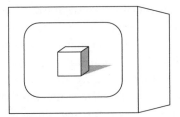

```
triangle(); rotate(); shadow(); color();
if (x<a) then z = 13; for i = 1 to 20; y = sqrt (z+10);
x → y; z → x; x * y → Z; A → x; y - x → z; z → B;
110010001101110111001000101101010000101010010010010001
```

Figure 7.5 Different realizations of a computer program. The upper panel shows a cube moving on the screen of the monitor. The lower panel indicates some realizations of the corresponding program; from top to bottom: object-oriented code, C code, assembly code, and binary code. (After Cruse et al. 1998b.)

walking, the network hypothesis predicted swing movements with small elevation. This prediction was tested experimentally. Contrary to the prediction, observations on real biological systems showed that elevation is fairly independent of step amplitude. This finding encouraged us to modify the hypothesis and to develop a new network demonstrating the required properties.

Application of this new network shows more realistic walking behavior in general. Curve walking, for instance, has been improved. This example shows that a simulation can have heuristic value, that is, it can stimulate new experiments. It also shows that simulation is not the final goal of scientific research; rather it is a tool leading toward further experimental investigations.

Differences between Natural and Artificial Systems

Let us now turn to a more general point. The examples discussed demonstrate that the simulations in this line of research are becoming more and more similar to the behavior of the real insect. Of course, the actual simulations are still far from being completely identical with the insect. Nevertheless, one may ask whether at the end of this process the simulation will become identical with the real insect or whether a fundamental difference will still remain. In other words, is there a fundamental borderline between artificial and natural systems?

This hypothetical borderline can be approached from two sides. As shown previously, artificial systems can become more and more similar to natural systems. But it also happens that natural systems can become similar to artificial ones: natural systems—in particular human beings—may become more and more artificial as artificial elements are applied to them, such as prostheses, artificial hearts, and even artificial sense organs like artificial cochleas or retinas. This is nothing new in principle. Even in Neolithic times, animals were changed artificially by the application of traditional breeding methods (which are now in the process of being supplanted by modern genetic technologies).

In fact the question is: are we speaking of two fundamentally different domains when we compare a simulation and "the real thing"? First of all, simulations—in particular hardware simulations— are real objects. Hardware simulations, it is true, differ from living objects. However, living objects—even different individuals of the same species—also differ from one another. So the question becomes: is there another, fundamental difference between simulation and a living object which does not exist between two living objects?

Intuitively, one may agree with the idea that in "simple" animals like insects the principles of information transformation, of dealing with the world and performing behaviors, may not fundamentally differ from those in artificial systems. However, Moravec (1988) has argued that this is also true for humans. He proposed the following gedankenexperiment: Assume that it is technically possible to replace one of the nerve cells of a human brain with an artificial neuron, which, to the necessary degree of exactness, shows the same input-output properties as the original neuron. Now imagine that this is done with a neuron in your brain. Do you think that you, your feelings, your self-experience, will be affected by this procedure? Moravec argues that there would be no such effect. If you have agreed with his view up to this point, Moravec goes on to propose the replacement of a second neuron. Still no change. A third neuron will be replaced in the same way—and so on, until your entire brain has been replaced by an artificial system. Even then, according to Moravec, your inner life will not have fundamentally changed. You still have an internal perspective and feelings. In this sense, you are still the same person. You have simply obtained a brain prosthesis, as you might have obtained an artificial heart. Of course, some superficial technical aspects of your life may change in both cases, but this does not change you in principle.[3] Concerning our question, we may now ask: Are you still a natural system, or have you become an artificial one? Have we crossed the borderline by following Moravec's procedure?

HIP Systems and NIP Systems

People may easily be misled when attempting to give intuitive answers to such questions. At conferences on artificial life (or sometimes at art exhibitions) one often sees demonstrations of animats. When observing the reaction of the audience, one quickly realizes that people (oneself included) are quite prepared to project feelings, intentions, or will onto these artificial creatures. This is the case even though we are rationally aware of the simple internal structure of these systems, which consist only of a few simple, reflex-like mechanisms. We readily fall into the "trap of subjectivity." This happens even more easily when the body of the robot is covered with fur, when it moves its head and looks around with big brown eyes. Apparently we have an innate animistic tendency, which projects the idea of a "self" onto any object as soon as it fulfills certain minimum conditions.[4]

Therefore it might be helpful to apply the following distinction, in which systems may be grouped into one of two domains. One group contains systems that have an internal perspective or subjective experience. Some may describe them as having a soul. Clearly humans belong to this group, along with, most probably, "higher" animals such as mammals or birds.[5] Systems in the other group do not have an inner life or a subjective experience. Most probably stones, wristwatches, and simple robots do not have an internal perspective. They have no subjective experience. Let us call systems with No Internal Perspective NIP systems, and systems that are able to Have an Internal Perspective (i.e., with subjective experience) HIP systems. It is a completely open question whether "simple" animals like insects belong to the HIP group or the NIP group, that is, whether or not they have a soul.

Using this distinction, an answer to the foregoing question—is there a fundamental difference between artificial and natural systems—may be easier to arrive at. Let us consider NIP systems first. In the case of an NIP system, I see no obstacle to a simulation coming closer and closer to the real system. Even if one insists that not only the principles of information transfer but also the protein-based biological structure are essential aspects of a natural system, this protein structure may in principle also be produced artificially. However, in my view a protein-based structure is not a necessary prerequisite. Two systems can be quite different in detail and yet show no difference with respect to certain essential properties. This may become clearer when, instead of comparing an animal and its corresponding animat, we compare two different animals. For the sake of simplifying the discussion, let us assume for a moment that insects are NIP systems.[6] Then a bumblebee may be considered a simulation (i.e., an incomplete version) of a honeybee (and vice versa), because many essential properties of the honeybee can be found in the bumblebee. Similarly, an artificial honeybee could possess the essential properties of a real honeybee. This means that, in the case of an NIP system, a simulation can provide a gradually better approximation—and therefore an explanation—of the functioning of the real, natural system. In this case, there is no fundamental borderline between natural and artificial systems. Both are part of a continuum. Therefore, for NIP systems, I see no limitation on the explanatory power of simulations, in particular hardware simulations.

With regard to HIP systems, the answer to this question is much more difficult. Again one might argue that, for example, a monkey may be considered a simulation (i.e., an incomplete version) of a human being (and vice versa). However, whether the argument of Moravec really holds for HIP systems—in other words, whether artificial systems can be built that have an internal perspective, a soul—remains an open question. Even if we were to construct such a system, how could we prove that it really has a subjective experience? Only the system itself (or should we then better say him- or herself?) can have this experience. For all other systems, only conclusions by analogy are possible, and these may be right or wrong, according to the possible flaws discussed previously.

So a cautious answer is that the question of whether a simulation can help to explain properties of subjective experience is still an open one. However, I believe that this is too pessimistic a view, and one that should not prevent us from attempting to perform such simulations (see, e.g., Cruse 1999 for a qualitative approach). These simulations will, for example, be supported by progress in research on the transition from sleeping to awake states and vice versa, because this transition describes the change between a state without and a state with an internal perspective. If we were to develop a sufficiently detailed model of these mechanisms, we might be more inclined to assume that simulations can also explain the phenomenon of subjective experience.

Conclusion

Software and hardware simulations (1) allow for concise description of large amounts of data, (2) provide hypothetical explanations, (3) allow for (testable) predictions, (4) allow us to test the feasibility of a hypothesis, and (5) may show emergent (i.e., unexpected) properties.

For NIP systems I see no limitations concerning the explanatory power of simulations. For HIP systems the situation is more difficult. Those properties related to the system's external perspective most probably can also be explained by simulations. Concerning its internal perspective, an explanation by means of simulation studies might or might not be possible.

NOTES

1. Why should we apply the term *intelligence* to such a "hardwired" system? A system may be called intelligent if it can find a new solution to a given problem. According to Lanz (2000), there are two types of intelligence: adverbial intelligence and nominal intelligence. Adverbial intelligence refers to problems that have already been solved and for which the solution might exist in the form of a dedicated system, such as the one described here for the control of multilegged walking. Nominal intelligence refers to problems for which a solution has not yet been found, but must be developed from scratch.

2. The term *dynamic* is used here in the physical sense, not as in the theory of dynamic systems, in which it is used in the sense of "dependent on time."

3. Based on these arguments, Moravec draws some exciting (if not horrifying) conclusions: you might produce a backup of your brain and thus a way to attain eternal life, or you might produce a copy of yourself in order to exist twice. With respect to transplantation, there is, of course, an important difference between heart and brain. If you obtain a heart transplanted from another person, you are still yourself. If you receive a brain transplanted from another person, you are now this other person.

4. This tendency may be helpful for a system acting in a complex (in particular, a social) environment.

5. Perhaps the whole universe is such a system. But we cannot see this because we are only small elements of it, just as my neurons do not realize that I have an internal perspective.

6. An assumption with which many people would agree, although personally I am not convinced.

REFERENCES

Brooks, R. A. 1995. "Intelligence without Reason." In L. Steels and R. Brooks, eds., *The Artificial Life Route to Artificial Intelligence: Building Embodied, Situated Agents.* Hillsdale, N.J.: Lawrence Erlbaum, 25–81.

Cruse, H. 1999. "Feeling Our Body—The Basis of Cognition?" *Evolution and Cognition* 5: 162–73.

Cruse, H., T. Kindermann, M. Schumm, J. Dean, and J. Schmitz. 1998a. "Walknet: A Biologically Inspired Network to Control Six-Legged Walking." *Neural Networks* 11: 1435–47.

Cruse, H., J. Dean, and H. Ritter. 1998b. *Die Erfindung der Intelligenz, oder können Ameisen denken?* Munich: C. H. Beck.

Dean, J. 1998. "Animats and What They Can Tell Us." *Trends in Cognitive Sciences* 2: 60–66.

Lanz, P. 2000. "The Concept of Intelligence in Psychology and Philosophy." In H. Cruse, H. Ritter, and J. Dean, eds., *Prerational Intelligence: Adaptive Behavior and Intelligent Systems without Symbols and Logic,* vol. 1. Dordrecht: Kluwer, 19–30.

Moravec, H. 1988. *Mind Children: The Future of Robots and Human Intelligence.* Cambridge, Mass.: Harvard University Press.

Müller-Wilm, U., J. Dean, H. Cruse, H. J. Weidemann, J. Eltze, and F. Pfeiffer. 1992. "Kinematic Model of Stick Insect as an Example of a 6-Legged Walking System." *Adaptive Behavior* 1: 155–69.

Steinkühler, U., and H. Cruse. 1998. "A Holistic Model for an Internal Representation to Control the Movement of a Manipulator with Redundant Degrees of Freedom." *Biological Cybernetics* 79: 457–66.

Wilson, S. W. 1991. "The Animat Path to AI." In J. A. Meyer and S. Wilson, eds., *Simulation of Adaptive Behavior: From Animals to Animats.* Cambridge, Mass.: MIT Press, 15–21.

8

The Semantic Challenge to Computational Neuroscience

Rick Grush

Department of Philosophy, University of California at San Diego

Some Distinctions

In discussing the relationship between the brain and computation, it will be helpful to begin with the following crucial distinction:

A   *Computation (or more felicitously a computer) as a tool for simulation*

Computers running the right software can be used to simulate a vast range of phenomena, from the antics of subatomic particles within a nucleus to the ways in which minor variations in conditions just following the Big Bang might affect the distribution of matter throughout the universe. One can simulate economic systems, human psychological performance, weather systems, and even computational systems themselves. What is important to note is that one can design and run computer simulations of a system or phenomenon without in any way taking a stand on whether or not that system or phenomenon is itself computing anything.

B   *Computation as a theoretical stance in cognitive neuroscience*

Many neuroscientists believe that one can shed light on the operation of biological neural systems by treating them as themselves carrying out computations. That is, unlike planets in the solar system, which *act in accord with various functions* but are not themselves computing those functions, the idea is that neural systems *really are computing* this or that function, *not merely acting in accord with it.*

With respect to B, we can make two further distinctions, depending on what we mean by *compute* or *computational*:

B1   *The brain (or parts thereof) implements some specific general-purpose computational architecture (e.g., a [finite-tape] Turing machine, or some other finite-state automaton).*

Few neuroscientists take this position at all seriously. Nobody really thinks the brain stores symbols corresponding to discrete ones and zeros and operates on them in a serial manner according to a state-transition table, and so forth. Nonetheless, it is useful to bring this position to center stage if only to be sure to avoid it.

B2   *The brain (or parts thereof) computes in the broad sense of implementing computable functions.*

The presumption is that by implementing such functions the brain processes information, and thus we might call this view the view that the brain is a *computational information processor* (more will be said about information processing later on). This is the serious contender, and most neuroscientists, especially those who call themselves computational neuroscientists, readily subscribe to it. In this broad sense, (implemented) Turing machines; finite- and combinatorial-state automata, including most (maybe all) connectionist networks; and many dynamical systems will count as things that compute.

In the remainder of this chapter, I take it as understood that all mention of computation is to be understood in the broad sense, and therefore that *computational neuroscience* is the field of study defined by the guiding assumption that biological neural systems are best understood as systems that process information by implementing computable functions.

This seems innocent and uncontroversial enough, but I try to show that such innocence is quickly lost. The trouble will come when one tries to keep B2 and A distinct, that is, when one tries to keep the notion of being a computational information processor (1) broad enough so that we can count things other than digital computers as computational information processors, while at the same time keeping it (2) narrow enough so that any system that can be simulated will not count as a computational information processor.

That all is potentially not well can be seen from the following passage from Churchland et al. (1990), an article that is considered one of (and perhaps the first and best of) the defining expressions of the project of computational neuroscience:

> A physical system is considered a computer when its states can be taken as representing states of some other system; that is, so long as someone sees an interpretation of its states in terms of a given algorithm. Thus a central feature of this characterization is that whether something is a computer has an interest-relative component, in the sense that it depends on whether someone has an interest in the device's abstract properties and in interpreting its states as representing states of something else. (48)

The first thing to note is the ambiguous use of the term *represent* in this passage. Phrases such as "representing states of some other system" suggest that what is crucial is some semantic relationship between a physical state of the brain and some other external object or state of affairs in the environment of the creature whose brain is under discussion. Let us call this sort of interpretation an environmental-semantic, or simply *e-semantic,* interpretation. On the other hand "interpretation of its states in terms of a given algorithm" suggests that the crucial semantic relationship is between a physical state of the brain and a variable, or value of a variable, of an abstract algorithm. Let us call this sort of interpretation an algorithm-semantic, or simply *a-semantic,* interpretation. One cannot help but notice the easy slide from e-semantics to a-semantics in the first sentence of the quote—and then back again in the last sentence. Though e-semantics is seldom far from the surface of the discussion by Churchland et al., by and large it is a-semantics that seems to be central. This gets spelled out just a bit further on:

> In a most general sense, we can consider a physical system as a computational system just in case there is an appropriate (revealing) mapping between some algorithm and associated physical variables. More exactly, a physical system computes a function $f(x)$ when there is (1) a mapping between the system's physical inputs and x, (2) a mapping between the system's physical outputs and y, such that (3) $f(x) = y$. (48)

This is clearly a-semantics, and it is difficult not to find this a bit unsettling. According to this proposal, everything is computing the function specified by equations that govern its physical behavior. Notice that exactly this constraint—the existence of a (revealing?) mapping between the physical states of some system and some algorithm (or computable function)—is both necessary and sufficient for being able to simulate that physical system by means of a computer.[1] That is, on the a-semantic account of what makes a physical system a computer, an account that is taken from the canonical defining expression of computational neuroscience, B2 collapses to A without remainder.

Perhaps this is what drove Churchland et al. to make the concession regarding the interest-relativity of computing. But these concessions do not solve the problem, they merely highlight it—for they imply that the only difference between A and B2 is that the B2 cases are the A cases such that someone cares enough about them to actually construct a computer simulation of them. We wanted to know when some physical thing, like the brain, really is a computer—really is computing. Churchland et al. seem to have simply given up at just the moment we needed their help most.

But I do not like to give up unless it is absolutely necessary. Perhaps we can get somewhere by making a further distinction, a distinction blurred in the expression of B2, between two senses in which B2 can be understood—a distinction that hinges on whether we are doing a-semantics or e-semantics:

B2a   *The brain (or parts thereof) computes in the sense that it implements computable mathematical functions.*

This is what we have just been discussing. According to this view, a system is a computer if there is a revealing mapping between its states and the input, output, and any "intermediate" variables of some computable function. The qualifier "revealing" need, and perhaps can, mean no more than that simulations exploiting that algorithm shed some light, for someone, on the operation of the system. This much will be true of any successful computer simulation.

B2e   *The brain (or parts thereof) computes in the sense that it processes information—it deals with what genuinely are information-carrying states—for example, states that carry information about objects or states of affairs in the environment.*

According to this view, it is e-semantics that is doing the work.

Notice that on the surely plausible assumption that the processing of information occurs according to tractable lawlike causal processes, the B2e cases are a subset of the B2a cases. That is, all B2e's will also be B2a's, since any tractable process is a B2a. The hope is that the same considerations that serve to differentiate the B2e's from the other B2a's will also serve to differentiate the systems that really are computing a function from those that are acting in accord with, but not computing, a function. In the next two sections I briefly present two case studies of computational neuroscience in action with an eye toward clarifying the distinction between B2a and B2e. The examples are a bit dated (each is drawn from the late 1980s), but that is not relevant. They have been chosen precisely because they have been taken to be parade-case examples of successful computational neuroscience, and because they bring out the difference between B2a and B2e.

Computational Neuroscience I: Koch

First an example at the level of the single neuron. Christof Koch (1990) reports studies, done in collaboration with the Paul Adams laboratory, on the bullfrog sympathetic ganglion cell:

We chose as the object of our study type "B" bullfrog sympathetic ganglion cells . . . for which the description of the various macroscopic currents is fairly complete. Due to their lack of dendrites, synapses are formed on or near the cell body and space clamp problems are absent. The aims of this work are (1) to separate the total ionic current into various distinct components, (2) to develop empirical equations that approximately describe the behavior of the currents under physiological conditions, and (3) to compare the computer simulations with cell responses during various experimental paradigms, for instance using pharmacological blockers. (107)

The equation used to model the neuron is

$$I = g_{Na}x^2_{Na}y_{Na}(V - E_{Na}) + g_{Ca}x_{Ca}y_{Ca}(V - E_{ca}) + (g_A x_A y_A + g_M x_M + g_K x^2 y_K + g_C x_C + g_{AHP}x^2_{AHP})(V - E_K) + g_{LEAK}(V - E_{LEAK}) + cdV/dt \tag{1}$$

What each of these variables means is not to the point. What is to the point is that this equation describes the physical processes that govern the relevant electrical behavior of the neuron. The equation is an abstract mathematical object, and the variable names have been chosen so as to suggest the names of the physical parameters to which they map (where *map* is clearly being used here in an a-semantic sense). For example, the variable g_{LEAK} is the a-semantic interpretation of membrane leak conductance.

The point of the investigations on which Koch reports is to determine whether or not equation (1) does in fact accurately describe the physical behavior of the neuron. This is tested by simulation studies, in which the behavior of a virtual neuron—whose behavior is guaranteed to match (1) because (1) is used to model it—is compared with the behavior of a real neuron under a variety of matched real and simulated conditions, including the influence of (real and simulated) pharmacological blockers.

The result of the simulations, of course, was that the behavior of the simulated neuron matched (to within some acceptable degree of accuracy) that of the real neuron under a variety of conditions, the match being close enough to merit some confidence that (1) was in fact the equation describing the physical behavior of that system.

This is a clear case of B2a. The only semantic relation appealed to is between states of the neuron and variables of an equation. There is no pretense to the effect that *information processing* is going on in the neuron, and hence there is no pretense to e-semantics. The neuron satisfies the Churchland et al. a-semantic requirement for performing a computation. But this is no surprise, since *any* tractable physical process satisfies this requirement. Of course, it may be the case that the neuron does process information because of the way it behaves, but if so, this fact is entirely external to and incidental to this simulation.

Computational Neuroscience II: Zipser and Andersen

One of the most fascinating and successful examples of computational neuroscience is the Zipser and Andersen (1988) model of the response properties of (some) neurons in posterior parietal area 7a.

First the initial biological data. Lesion data had long suggested that this cortical area is crucial for spatial representation, but little was known about how this was accomplished. Single-cell recordings in the early and mid-1980s (e.g., Andersen et al. 1987) showed that there are at least three groups of cells in this area: some have responses tied to the retinal location of the projection of a visually presented stimulus; a second group responds to the orientation of the eye in the head; and a third group shows responses that are influenced both by the location of a visual stimulus on the retina (retinal location) and by the orbital orientation of the eye. Such cells fire maximally given only a certain combination of retinal location and orbital position.

Next the simulation. Zipser and Andersen (1988) designed a connectionist network whose inputs were (1) the location on the retina of a given stimulation and (2) the orientation of a simplified two-dimensional eye. The network was trained to learn the direction of the stimulus relative to the "head" given these two pieces of information. The result was that the model learned the task, meaning that it learned to provide the correct output given the two inputs. Furthermore, it was discovered that the response profiles of "hidden" units of the network had properties that seemed to be similar to the properties of actual neurons in area 7a.

The payoff, however, was that the connectionist network could be analyzed in great detail to determine exactly how it was solving the problem. It was learned that the hidden units were implementing planar gain fields, in which each unit had a gaussian retinal receptive field, but in addition had its activity modulated linearly by the orientation of the eye. In effect, a given unit would fire most strongly with a given combination of retinal stimulus location and eye orientation. (It was also determined that the same mechanisms could be used by a connectionist model to compute body-centered locations, by having hidden units that acted as planar gain fields combining retinal location, eye position, and head orientation. Subsequent single-cell recordings confirmed the existence of cells with such response profiles.)

The situation with this simulation is more complex than the situation with the last. In this case, as in the last, there is an a-semantic relation between physical states of neurons and variables of a mathematical function. That is, we can suppose that the function computed by an idealized posterior parietal neuron is something like $f = (\vartheta - \vartheta_p)$ $g(r - r_o)$, where f is the a-semantic interpretation of firing frequency, $\vartheta - \vartheta_p$ is the interpretation of firing frequency of one presynaptic neuron, and $g(r - r_o)$ is the interpretation of firing frequency of the other presynaptic neuron. This is a B2a explanation: the state of the cell is some function of the states of the cells that synapse on it.

But in addition to this a-semantic mapping, there is an e-semantic mapping: f is the e-semantic interpretation of stimulus distance from preferred direction relative to the head (high firing rate means low distance), $\vartheta - \vartheta_p$ is the interpretation of the difference between the actual and preferred eye orientation, and $g(r - r_o)$ is the interpretation of (a gaussian of) the distance from the retinal location of stimulation from the receptive field. This provides for a B2e explanation: the cell is processing information about retinal location and eye orientation to provide information about the direction of stimulus relative to the head.

Representation

We have before us two examples of computational neuroscience in action. The study by Koch involving the bullfrog ganglion cell appears to involve no more than would be involved in any example of a computer-simulation-cum-experimental-testing endeavor of the sort familiar in computational physics, economics, meteorology, and dozens of other areas. This is by no means to say that it is uninteresting. Quite the opposite: such interaction of experiment and computer simulation has proven to be an incredibly powerful tool for shedding light on the operation of complex systems whose principles would otherwise remain hidden. The point is that this model is a clear case of B2a, and B2a is just a tarted-up version of A.

The Zipser and Andersen model (and subsequent experiments based on it), however, seems to be a good example of a B2e. It sheds light on how a set of neurons interact so as to *represent directions of stimuli in egocentric space.* Of course, one *can* describe the behavior of the neurons in the posterior parietal in a thin B2a manner, as discussed in the previous section. But one also has the B2e description available. One plausible suggestion is that it is the availability of such an alternative, *representational,* description that sets B2e cases apart from B2a cases. This is of course what one would expect, since B2e cases are just cases in which there is a semantic relation established between the neural states and something in the environment. This something will be what the neural state represents. (The question of what it is that makes this representational description available and appropriate— what establishes the semantic relation—will be taken up shortly.)

So, if there were some principled means to determine which states are representing aspects of the environment, we could exploit this to determine which of the B2a cases are in fact B2e cases. We would have the means to distinguish those systems that are genuinely computational in the required sense, and there would be no danger of computational neuroscience's being assimilated without residue into the general category of computer simulation studies.[2]

I will assume that a bald interpretationalist stance here is unacceptable. That is, I am assuming—as should any self-respecting computational neuroscientist—that there is a fact of the matter concerning whether or not a given neural state is representing something: a fact that is neither created nor imperiled by the interpretive whims of whoever might be examining the neural system. But the conviction to the effect that there is such a fact of the matter is not enough. What we need is some means to explain why it is that *this* state does, but *that* state does not, really represent something. If we lack such means, then computational neuroscience will have to proceed on faith and in ignorance.

Now I—like many people who spend time worrying about issues of representational content—happen to think that such means can be provided. However, the theories about how this happens that have the most currency are, I think, unworkable. This is not the time to go into a rather involved positive account of my own (for this, see Grush in preparation). But this is the time to show why neither of the two most popular and initially plausible accounts will work.

Informational Semantics

A groundbreaking but ultimately untenable attempt to show how some things are intrinsically representational is the informational semantic approach, whose locus classicus is Dretske (1981).[3] According to this approach, one can appeal to the information carried by the state of some system in order to fix the content carried by that system. For instance, suppose that a given neuron in my cortex fires at a rate faster than 50 Hz when and only when there is a square in my visual field. If this is the case, then one can say that that neuron's state of firing over 50 Hz carries information to the effect that there is a square stimulus in my visual field (and also that the neuron's state of firing *under* 50 Hz carries information to the effect that there is *not* a square stimulus in the visual field). Anyone familiar with the neurophysiology literature on perception, cognitive mapping, motor control, and related fields will recognize that this approach is widely assumed by neuroscientists, who look for causal covariation to establish that something (e.g., a moving bar, a face, a twitching paw, a "place") has representational content.

The problem with this suggestion is that information is as ubiquitous as are law-governed causal processes. The temperature of the outer surface of my coffee cup is a well-behaved function of the following: the original temperature of the coffee, amount of coffee, original temperature of the cup, time since the coffee was poured, ambient temperature, and heat-transfer properties of the cup. Given that all parameters but the room's ambient temperature were fixed when the coffee was poured, the cup is processing information about the ambient temperature of the room. In general, any physical state of anything carries information about the states of other physical entities that enter into causal dialogue with that state, and hence it will be possible to view any causally fertile (law-governed) process involving the evolution of physical states *of anything* as information-carrying. (Some examples of ersatz information-carrying in the brain are provided in the next section.)

As it happens, problems such as these and many others have led to the collapse of informational semantics, its prevalent role in the pretheoretical intuitions of neuroscientists notwithstanding. Evidently we must look elsewhere for our semantic salvation.

Biosemantics

Yet another proposal, one that is increasingly popular, is appeal to biological functions (see Millikan 1984 for the locus classicus; see also Millikan 1989). According to this proposal, the proper function of *d* in *S* is to detect *c* iff *d* was replicated because it carried information about *c*; and for our purposes we can take *detect* to have the force of *represent*.[4] In short, the idea is that the cell in my posterior parietal cortex (PPC) represents the direction of a visual stimulus by virtue of the fact that that PPC neuron (or more plausibly the architecture of the PPC that induces such neurons) was replicated because it carried information about the direction of a stimulus relative to the head.

How does this solve the problem that bedeviled informational semantics? Consider the following example. The skin on the bottoms of my feet has cells that respond to pressure. Given this, and other things being equal, the more I eat, the more these cells will fire, since the more I eat, the more I will weigh, and the more I will weigh, the more pressure will be applied to the bottoms of my feet. Furthermore, there are cells in my primary somatosensory cortex that are responsive to the activity of these pressure-sensitive cells. So in terms of carrying information, these cells in my somatosensory cortex carry information about how much I have eaten recently. This is just another example of the ubiquity of information.

But according to the biosemantic account, those cells are not representing how much I have eaten recently, because the response profiles of these cells were *not* replicated because they carried information about my eating behavior. The explanation of their replication would rely on other considerations altogether, having to do in part with detecting features of the surface with which the feet are in contact. There are other sets of neural structures that were replicated because the information they carried was useful for—*and actually used for*—the regulation of eating.

So the biosemantic gambit is this: with a prior notion of some biological function in mind, such as facilitating accurate motor control, one can view certain physical states of the organism as contributing to this function because they carry information that is crucial for the appropriate execution of that function. Information about where something is is crucial for my ability to grasp it, for example, whereas the information in the soles of my feet about how much I ate for dinner, although genuinely informational, is not serving any function (though it could be incorporated into a system that does: imagine a creature whose foraging behaviors were triggered by the pressure sensed on the bottoms of its feet, so that as the creature became lighter it experienced a greater drive to find and eat food). Complex biological systems have components which are such that an explanation of why those structures are the way they are appeals to the fact that their carrying certain sorts of information facilitated their evolutionary selection. This will be true of only a small subset of such information-passing states.

This sort of account has wide and expanding appeal within the philosophy of psychology literature on content. But I now show that, its appeal notwithstanding, it is not a proposal that the defender of computational neuroscience can embrace without paying a rather high price.

Physical systems, qua systems describable by the laws of physics, are state-determined systems. This means that the future behavior of the system is determined (perhaps statistically determined) by its current state. If a ball is at location x_1 with momentum y at t_1, then, in the absence of any forces acting on it, it will be at location x_2 at t_2. What is crucial to note is that the laws of physics *do not care* how the ball got to location x_1, or how it came to have momentum y. One will not find different laws depending on whether the ball was struck by another ball or was pushed by a hand or magnetic field or was expelled a billion years prior to t_1 from a nova. How it got into that state is irrelevant. All that is relevant to its behavior—all one needs in order to apply the laws of physics and determine what the ball will do—is its *current* state (and the forces currently acting on it, if any).

If we leave the realm of physics momentarily, we can see that it is often the case that things other than an object's narrowly physical state can be crucial to its identity and status. For example, whether a dollar bill is genuine or counterfeit is not determined by its physical state. It is determined by who printed it and where. Two pieces of paper might be in identical physical states—molecule-for-molecule identical—and yet it can nevertheless be true that one is a genuine dollar bill and the other is not. Now we can always decide to include anything we want in a description of something's state. That is, we can expand our notion of a *state* to include more information about it than is required for application of the laws of physics. We might, for example, decide to include a new variable in the state of all objects that indicates their place of origin—perhaps a variable T that takes the form of a binary +/−, so that an object is $T+$ if it was printed by the U.S. Treasury Department and $T-$ otherwise.

If we expand our conception of a thing's current state in this way, then we can make it the case that whether or not a piece of paper is a genuine dollar bill can be determined by its current state. And given that locations and processes of paper production and printing are physical things and processes, we might not feel *too* uncomfortable in continuing to call this expanded state description a physical state description. But this should not divert attention from the fact that an object's T value is unlike its mass, momentum, and position in a number of important respects. For example, I can take a piece of paper into a laboratory and determine its mass, momentum, and other properties. But if I want to determine its T value, the physicist and chemist are useless. I will need to hire a private investigator to try to track the history and origin of the piece of paper.

Now, according to the biosemantic account, whether or not a state is a representation is analogous to the question of whether or not a piece of paper is a genuine dollar bill. It is a function of the history of those mechanisms that support that state and not a function of the current (narrowly) physical state of the brain. To illustrate: suppose that tomorrow a bolt of lightning strikes a swamp, and the resultant release of energy combined with the chemical ingredients at hand produces an exactly identical molecule-for-molecule replica of you—

call it *swamp-you*. (Let us suppose for the sake of example that, just like you, swamp-you is created with a book in hand, and is looking at a page just like the one you are currently looking at.) The odds of this actually happening are entirely irrelevant.

Now let us suppose that you are looking at the word *swamp* at exactly the moment when swamp-you is created, and in fact swamp-you's eyes are also directed at an inscription of the word *swamp*. Now if biosemantics is correct, then some cell in your cortex is firing because it is representing a line with a certain orientation, and this is part of what allows you to recognize the word *swamp*. However, according to the biosemantic proposal, swamp-you—although it is in a molecule-for-molecule identical state to yours when you are perceiving the word *swamp*—is not perceiving anything, not representing anything, not thinking anything. Swamp-you's "head" quickly swivels around and produces a sound that we would categorize as an utterance of the form "How the hell did I get in a swamp?" (or whatever you would produce if you suddenly found yourself in a swamp while in the midst of reading a book). But, according to the biosemantic proposal, appearances notwithstanding, swamp-you is not thinking anything, not perceiving anything, because not representing anything.

Why? For the simple reason that in such a case the explanation of why a certain cell in swamp-you fires in a certain way would make *no appeal to evolutionary pressures*, since this swamp-you has *no evolutionary history*. Be careful not to be taken in by the fact that swamp-you is just like something that *does* have an evolutionary history. This is not relevant. (Being *just like* something that is worth $20 does not make counterfeit $20s really worth $20.) The explanation of why such and such a potential is present on this membrane (i.e., why this cell is firing) would appeal to physical events immediately preceding the presence of the potential, and an explanation of why the physical system was structured so as to have that sort of effect would appeal to the contents and configuration of the goo in the swamp before the lightning struck and the details of what happened upon the strike. Given this, it follows that the biosemantic account swamp-you is not really representing anything, since according to the biosemantic account something is a representation only by virtue of being the product of the correct evolutionary selection pressures. And, on the other hand, the explanation of why you are representing such things is not given by the details of what is happening in your brain.

Many people will take this to be a *reductio* of the evolutionary-teleosemantic account, as do I. Nevertheless, the counterintuitive conclusion has been embraced by proponents of this account—an admirable example of bullet-biting at the very least. My goal for now is not to provide any (more) reason to abandon biosemantics, but rather to point out that, whether or not biosemantics is correct, its adoption provides no comfort to the computational neuroscientist.

Simply put, evolutionary history is not something that shows up in the occurrent causal operation of the brain, any more than a dollar bill's T value shows up in its occurrent causal dealings. Note that all the computational neuroscience in heaven and earth would fail to detect a difference between swamp-you's brain and yours—they are, after all, molecule-for-molecule identical. Yet according to the biosemantic account, your brain, but not that of swamp-you, is really computing and representing; you, but not swamp-you, have a mind. So if biosemantics is correct, then the following questions are *not* among those that computational neuroscience can answer *or even address:* Are cells in your visual cortex representing things in the environment? Are cells in your hippocampus representing locations in allocentric space? What is the relation between the brain and the mind? And many more. You get the idea. In effect, every question that cognitive neuroscience ought to be able to answer if in fact it is the physical structure of the brain that explains thought, cognition, imagination, and perception will turn out to be a question beyond the reach of cognitive neuroscience and computational neuroscience. If biosemantics is correct, we must consult not the neuroscientist in order to determine if the cell in your brain is representing a face or a line or a color; rather we must hire a private investigator.

I now address two objections to what I have just said. The first is that the conclusion I draw is not as bad as I make it out to be, it just means that we must be interdisciplinary in our approach to cognition by including some evolutionary biology. The second is that examples like swamp-you are irrelevant to actual science, their ability to fascinate metaphysically inclined philosophers notwithstanding.

Objection 1   *OK, fine. So neuroscience by itself can't tell us which things can really represent, and hence can't tell us which things are really computational. So we enlist the aid of the evolutionary biologist. And we still have a perfectly legitimate scientific enterprise. What's the big stink?*

Reply 1. If it were this easy, then there would be no stink. But it's not this easy. First point. This is not the sort of innocent appeal to interdisciplinary study that is rather popular in cognitive science. The innocent variety is motivated because all current windows on the brain and cognition suffer from technical limitations—we cannot make single-cell recordings of every cell in a human brain while it is doing math, for example. And since each technique (e.g., single-cell recordings, positron emission tomography, evoked response potential, evolutionary biology) supplies some constraints, we will do well to take advantage of those constraints. But the biosemantic appeal to evolutionary biology is different. *It would remain even if we knew all the occurrent physical-causal things there were to know about the brain in question.* Any self-respecting physicalist should find this an unacceptable pill to swallow.

Second point. Evolutionary biology is of exactly no help here anyway, because it begs all the crucial questions. Specifically, the evolutionary biologist is not in the business of taking some animal and performing experiments on it, or observing it, or whatever, in order to tell whether it is (1) a real product of biological evolution or (2) a molecule-for-molecule replica of something that perhaps was—a swamp-animal, so to speak. The evolutionary biologist simply assumes that there are no swamp-animals running about, and with that reasonable but unverified assumption in hand proceeds to determine what were the most likely lines of evolution, series of adaptations, and so forth in the history of the species of which this animal is, *presumably,* a member.

Objection 2   *This swamp-man mumbo-jumbo is fine for metaphysically inclined philosophers, but neither philosophers who understand science, nor scientists themselves, should take it seriously. Such a creature would be so improbable that we need not take its possibility seriously. And even if one were to appear, we would have a misconception about that one particular individual, but this would leave the rest of our science completely unaffected.*

Reply 2. This objection simply fails to recognize the nature of the problem that swamp-you creates. The problem is that if we expect neuroscientific explanations of things like consciousness, representation, psychology, and the rest to be forthcoming, then swamp-people—molecule-for-molecule (and hence neuron-for-neuron) replicas *without* a psychology, consciousness, or representations—should not be possible *at all*. This is entirely parallel to saying that if special relativity is true, then massive particles being accelerated past the speed of light are not possible *at all*. I take it that the following line of reasoning would be scientifically disreputable: "Jones has shown convincingly that if we were to apply 10^{80} joules of energy appropriately to this dime, it would have speed $c + s$ (for some positive finite speed s). But this should not worry those of us who believe in relativity. The odds of anything ever having that much energy directed to its motion are almost nil, and in point of fact on our planet this simply never occurs. Perhaps wild-eyed philosophers should worry, but no real physicists should take this at all seriously." I take it that this would not be an intellectually reputable thing for a scientist, or anyone for that matter, to say. It might be reputable to deny the inference, claiming or arguing that in fact 10^{80} joules *would not* accelerate anything past c. That would be fine. What is *not* fine is to accept the antecedent, to accept the consequence that 10^{80} joules would accelerate it past c, and then act as though its low likelihood of occurring, or the fact that around here it simply does not in fact occur, or that even if it did occur it would just mean that our theory got *one* out of billions of force-acceleration explanations wrong, somehow renders this fact unproblematic to the acceptability of our apparatus of physical explanation—an explanatory apparatus that has relativity as a central core.

In *exactly* the same way, it is simply intellectually disreputable to maintain both biosemantics (which entails that swamp-people are possible) *and* the claim that neuroscience will be able to explain consciousness, representation, and psychology. Now, of course, it seems to me that there is an obvious and easy option that presents itself here: *ditch biosemantics, pronto*. With all apologies to Millikan and company, I think we should bronze it, polish it, and stick it in the museum of brilliant, noble, admirable, yet ultimately incorrect theories of which the history of philosophy boasts.[5] Then we can accept the reasonable premise that, in the extremely unlikely event that a bolt of lightning were to create a molecule-for-molecule replica of you, it would represent and be conscious, just like you. And we can also accept the idea that explanations for cognitive phenomena will be, eventually, produced by cognitive neuroscience. The price that neuroscience pays for ditching biosemantics is that, in doing so, it ditches one attempt to solve the problem of determining which things really are computational and which are not. But neuroscience was getting the short end of that stick anyway. It solved the problem only by denying neuroscience any say in the matter.

Conclusion

The thrust of this chapter has been negative. I have tried to show that, given the currently available theoretical tools, one cannot mark out computational neuroscience as an enterprise whose job it is to explain the neural basis of cognition. What is missing is the ability of computational neuroscience to provide an explanation—a theory-internal explanation—of why some neural states are representations and what the semantic import is of those neural states that are representations. Informational semantics fails to distinguish any useful subclass of states as representations, and biosemantics (as with any externalist content-assignation scheme), even if it works, does so only by depriving computational neuroscience per se of any claim to be able to explain cognition. If my analysis is correct, what is needed is an internalist semantics—a theory that can explain, given only appeal to mechanisms internal to the brain itself, why some of those states have semantic import and what that import is. I think such a theory can be formulated, but my defense of this claim will have to wait for another day (see Grush in preparation).

NOTES

I thank Peter Machamer and Gualtiero Piccinini for feedback on an earlier version of this chapter and participants in the Fifth Pittsburgh-Konstanz Colloquium in the Philosophy of Science for discussion of a still earlier version.
1. That it is necessary can be seen by noting that implementing algorithms is just what computers do. If there is no algorithm, then one cannot implement it on a computer. Of course, one may not know the exact algorithm before one does the simulation. One might find out what the algorithm is by training a connectionist network, for instance. The sufficiency follows by definition. Given that computers are general-purpose (one may need to make idealizations about amount of memory here, of course), meaning that they can execute any computable function, having a computable function is sufficient for being able to implement it on a general-purpose computer.
2. Though I arrive at them by different means, my conclusions here are similar to those of Shagrir (1994).
3. In what follows, I provide only the roughest sketch of a proposal that has had many clever defenders, who have provided responses to the sorts of concerns I

raise. I am not here trying to kill informational semantics, but merely recounting, in caricature form, some of the main sources of pressure that led to its death.

4. As was the case with informational semantics, my gloss here ignores a number of central refinements and features of Millikan's proposal, but those refinements and features are inessential to my purpose.

5. I hasten to admit that, aside from pointing to the counterintuitive example of swamp-people, I have provided no *argument* here for rejecting biosemantics. I am making a recommendation. But I expect it to be convincing only upon the provision of some arguments, and preferably a good alternate account. The alternative account will be provided in Grush (in preparation).

REFERENCES

Andersen, R. A., R. M. Siegal, and G. K. Essick. 1987. "Neurons of Area 7 Activated by Both Visual Stimuli and Oculomotor Behavior." *Experimental Brain Research* 67: 316–22.

Churchland, P. S., C. Koch, and T. Sejnowski. 1990. "What Is Computational Neuroscience?" In E. L. Schwartz, ed., *Computational Neuroscience.* Cambridge, Mass.: MIT Press, 46–55.

Dretske, F. 1981. *Knowledge and the Flow of Information.* Cambridge, Mass.: MIT Press.

Grush, R. 1997. "The Architecture of Representation." *Philosophical Psychology* 10(1): 5–23.

———. In preparation. *The Machinery of Mindedness.*

Koch, C. 1990. "Biophysics of Computation: Toward the Mechanisms Underlying Information Processing in Single Neurons." In E. L. Schwartz, ed., *Computational Neuroscience.* Cambridge, Mass.: MIT Press, 97–113.

Millikan, R. 1984. *Language, Thought, and Other Biological Categories.* Cambridge, Mass.: MIT Press.

———. 1989. "Biosemantics." *Journal of Philosophy* 86(6): 281–97.

Shagrir, O. 1994. "Computation and Its Relevance to Cognition: An Essay on the Foundations of Cognitive Science." Unpublished Ph.D. dissertation, Department of Philosophy, University of California, San Diego.

Zipser, D., and R. A. Andersen. 1988. "A Back-Propagation Programmed Network That Simulates Response Properties of a Subset of Posterior Parietal Neurons." *Nature* 331(6158): 679–84.

9

Functional Imaging Evidence

Some Epistemic Hot Spots

James Bogen
Pitzer College, Claremont, California

Why Epistemologists Should Think about Functional Imaging

Functional imaging (FI) research employs PET (positron emission tomography) and fMRI (functional magnetic resonance imaging) technologies to study the working brain. FI can do what other techniques cannot. Unlike standard cognitive psychology experiments, FI provides evidence from which to estimate levels of neuronal activity involved in the performances of cognitive tasks. Unlike electro-encephalographic and related techniques, FI can indicate the anatomical locations of functionally significant neuronal activity. Unlike single-cell recording, it is minimally invasive. Unlike neurocognitive techniques for studying the deficits of brain-damaged subjects, FI can study normal brains. Such attractions have encouraged the widespread use of PET and fMRI. The growing importance of FI research is one good reason for epistemologists to interest themselves in peculiarities of this methodology such as those I discuss in this chapter.

A second reason is that functional images fit empiricist accounts of evidence so poorly that they pose problems for empiricism's friends and provide ammunition for its foes. Old-fashioned empiricists thought the test of a scientific claim to be how well it stacked up against evidence delivered by the senses—whether unaided or aided by magnifying or amplifying devices. Their views were challenged by evidence produced by equipment that—instead of helping the senses discern what would otherwise be too faint, far away, or small for them (that is what stethoscopes, telescopes, and magnifying glasses do)— replaces the senses as sources of evidence (as do Geiger counters and galvanometers). Our sensory systems are simply not equipped to pick up and perceptually discriminate the features of radioactivity and electrical currents to which these instruments are attuned. It is not too much of a stretch for empiricists to meet this challenge by extending their notion of empirical evidence to apply to what registers on observational equipment instead of human sensory systems. Functional images are examples of a kind of evidence that presents a different and much more serious challenge. As we will see, this is because statistical and other sorts of mathematical analysis figure in their production in a way that blurs the distinction between producing empirical data and interpreting the data in order to draw conclusions about a system under investigation.[1] Unlike the distinction between what can be perceived by humans and what can register only on experimental equipment, the distinction between producing and interpreting data is far too central to empiricism for its adherents to give up.

Two Questions Functional Images Are Used to Answer

Functional images look like colored pictures of the brain. The colors represent computed levels of biological indicators (BIs), which vary with neuronal activity. Reds are often used for high BI levels, violets for low levels, and oranges through blues for intermediate levels. A color's location in the image indicates the anatomical location of the BI it represents.

PET subjects are injected with water or made to breathe a gas containing ^{15}O, a radioactive oxygen isotope that emits positrons as it decays. Having entered the blood, ^{15}O is carried to the brain. Blood vessels expand to raise blood volumes in neuronally (electrochemically) active regions of the brain. Because positron emission varies with the volume of radioactively marked blood, the products of ^{15}O decay can be used to measure neuronal activity in a given area of the brain. (An older technique, now seldom used, relies on positrons emitted by ^{18}F from radioactively tagged sugar molecules, which accumulates in nerve cells as a metabolic by-product of neuronal activity; Corbetta 1998, 96.)[2] fMRI technology exploits magnetic field effects to produce radio signals, from which relative amounts of oxygenated hemoglobin in venous capillaries can be calculated.[3] These quantities can be used to study neuronal activity because they vary with blood levels, and hence with electrochemical activity in nearby neurons. Functional images are formed from positron or radio signals emitted from subjects' brains at rest or during the performance of a mental task.[4] The colors of the resulting visual display give rise to the popular description of functional images as showing where the brain "lights up" during the performance of a mental task.

FI research typically assumes

A1   *BI cognitive significance.*

"The performance of any task places specific information processing demands on the brain. These demands are met through changes in neuronal activity in various functional areas of the brain" (Petersen and Fiez 1993, 513).

A2   *Decomposition and distribution.*

The performance of a task has as its components simpler functions distributed throughout different brain areas. None of these areas is responsible for more than part of a task of any complexity. As Petersen and Fiez (1993, 513) remark, "there is no tennis forehand area."

A3   *Localization.*

Each of these areas makes a specific contribution to the performance of the complex task, performing a function that can also belong to the performance of other complex tasks (Petersen and Fiez 1993, 513–14).

If these assumptions are correct, functional images can be used in arguments for "the identification of specific sets of computations with particular areas of the brain" (Petersen and Fiez 1993, 514).

Petersen and Fiez (1993, 515, 516) go on to point out that functional images are typically used to answer the following questions:

Q1   *What are the anatomical regions of the brain whose neuronal activity is peculiarly, or at least closely, associated with a given task?*

Q2   *Which information processing functions (which relatively simple components of a complex task) are associated with the neuronal activity that has been found to occur in a given anatomical region of the brain?*

For example, functional images suggested that there was activity in the left dorsolateral prefrontal cortex during performances of a number of different tasks, including verb generation, silent monitoring of a list of words for members of a specified semantic category, and passively viewing words and pseudo-words. No such activity appeared during silent monitoring of false fonts (strings of marks that are not letters of the alphabet). The question of which (if any) computations common to the first three (but not the last) of these tasks is carried out in the left dorsolateral prefrontal cortex is an instance of Q2.[5]

A Received Worry about Functional Image Methodology

Some epistemically troubling peculiarities of FI research arise in connection with its use of subtraction images. Ignoring details, these visual displays are produced by the application of elaborate computational procedures (this is what is meant by "subtracting") to the numerically coded versions of two images. The computations are designed to purge one image of unwanted features belonging to the other. "Compound subtraction" images are produced by subtracting the image of a brain performing what is believed to be (or at least to approximate) a relatively simple component of a task from the image of a brain performing the task whose component it is thought to be. In "simple subtraction," baseline images of a resting brain are subtracted from images of a brain at work on some task (Sidtis et al. forthcoming). Subtraction images are used to correct for errors arising from the fact that functional images of a brain engaged in a complex cognitive task inevitably exhibit signs of neuronal activity of no interest to the investigator. Brains can work on a number of unrelated tasks at the same time. While performing the cognitive task the investigator wants to study, the subject may observe and think about the laboratory decor, try to recall a phone number, and hear Ben Harper singing "Susie Blue" in her head. Furthermore, according to A1–A3, a single image of a brain at work on a complex task should display the BIs of a number of anatomical regions engaged in its component subtasks. To study any given component, the investigator must distinguish BIs associated with it from BIs associated with other subtasks. Hence the need for subtraction images. I believe the most interesting epistemological problems posed by subtraction images are instances of some more general problems arising from the computations that go into their production. Other writers have been more concerned with difficulties arising from the role of assumptions from cognitive psychology in the production of subtraction images. I now consider such difficulties. I believe they are not as serious as they are claimed to be. After explaining why, I begin sketching what I take to be FI's most epistemically intriguing peculiarities in the next section.

Simple subtraction assumes the investigator can identify the conditions under which neuronal activity will be at a baseline rate. Compound subtraction assumes the investigator has correctly identified the subtask components of the relevant complex task. It assumes further that there is no significant difference in level or location of neuronal activities associated with subtasks performed in isolation from one another and together in the course of performing the complex task. Thus subtractions assume the correctness of models from cognitive psychology; critics of FI have seized on this assumption. In his highly critical review of some FI research on language processing, David Poeppel (1996, 317, 341–42) complains of "insufficiently detailed," uninformed, and unsubstantiated "intuitive" ideas about tasks and their components. He also objects that tasks will not divide up into components that can be "subtracted from one another yielding a component subprocess" unless they "are processed in serially ordered, discrete stages" and there are no feedback mechanisms by which components of a complex task can modify one another. He claims that the first of these assumptions is probably false, and he believes that there is substantial evidence to argue against the second (Poeppel 1996, 344). Philosophers of science took up such issues at a 1996 Philosophy of Science Association symposium at which van Orden and Paap agreed with Poeppel. They also suggested that the acceptability of subtraction images as evidence depends upon the correctness of the relevant cognitive psychological assumptions to such an extent that arguments based on these images are bound to be viciously circular (van Orden and Paap 1997, S89). In the same symposium Stufflebeam and Bechtel (1997, S99) take a brighter view of FI but find its reliance on cognitive models and its assumption that "additive, independent processes can be inserted or deleted separately into a cognitive process without altering others in the process" troublesome.

One thing to say about all of this is that subtraction is not always necessary. So many fMRI images can be produced in such a short time that BI levels associated with some tasks can be isolated simply by alternating tasks and scanning the brain as it works first on one, and then on the other (Bavelier et al. 1997, 681). More importantly, as illustrated by the following example from the literature Poeppel reviewed, the cognitive psychological commitments involved in subtraction may be quite unproblematic.

Petersen and others conducted FI experiments that bear on Alan Baddeley's hypothesis that the short-term memory mechanisms at work in a number of language tasks include a "phonological loop" consisting of "a phonological store that is capable of holding speech-based information, and an articulatory control process based on inner speech which is, among other things, capable of taking written material, converting it into a phonological code, and registering it in the phonological store" (Baddeley 1998, 52–53).

Petersen knew from earlier PET experiments that a region in the superior temporal gyrus (STG) that appeared to be relatively inactive during passive viewing of a blank screen appeared to be active during passive listening to words read aloud. Some of the passive listening STG activity fell "outside of the primary auditory cortex" in an area in which significantly high BI levels appeared during the performance of linguistic tasks. Petersen et al. asked whether "(h) this part of the STG performs phonological processing" or "stores common internal codes for auditory and visual words taking the form of a sound or a name code for the word" (Démonet et al. 1996, 370). Evidence for either disjunct would be congenial to Baddeley's proposal. To test (h) they made PET images of subjects whose task was to find rhymes among visually displayed pairs of words and nonwords. The investigators subtracted images from other conditions—including the passive viewing of displayed pairs of words and nonwords, and passive fixation on a blank screen—from the rhyme task images. Although the rhyming task stimuli are visual rather than auditory, the subtraction images indicated activity in the area of STG outside the primary auditory cortex that had been active during passive listening to words in the earlier experiments. Subtraction images from a second group of experiments indicated no activity above baseline when subjects performed linguistic tasks (e.g., "generating verbs in response to visually presented nouns" and reading without responding to visually presented words) that do not depend, as does the rhyming task, on how the displayed words would sound if pronounced (Démonet et al. 1996, 371).[6]

These results count in favor of (h) and are as they should be if the relevant part of the STG performs one or both of the functions of Baddeley's phonological loop (Démonet et al. 1996, 366; Poeppel

1996, 382). The reliability of these experiments as tests of (h) and their bearing on Baddeley's hypothesis do not depend upon the accuracy of precisely detailed psychological assumptions. It is assumed, for example, that neither reading nor passively viewing words is a component of passive listening, that listening to words is not a component of passively viewing a blank screen, that hearing and processing spoken words are not parts of the rhyming task (its stimuli are all visual), that rhyme identification is not a component of reading or verb generation, and that the sounds of words are irrelevant to the performance of these tasks. Such decompositional assumptions are hardly tendentious. Petersen's arguments also assume that the cognitively significant neuronal activities are not too diffusely distributed for functional images to have any evidentiary bearing on (h). But this assumption is well supported by the spatial separations of displayed BIs and by evidence from experiments that do not involve functional imaging. Finally, worries about feedback can be checked by comparing the functional images, including subtraction images, for the various tasks Petersen et al. used. Poeppel and the others are certainly right to say that decompositional models figure in the interpretation of subtraction images and the choice of subtraction tasks. But experimentalists usually go out of their way to design experiments whose results will not be hostage to uncheckable or problematic cognitive assumptions. Our example is typical of how well the best FI experimental designs can succeed in this.

What I take to be the most interesting epistemological questions about FI arise from other factors. Some of these factors are inherent in the processes by which PET and fMRI images are produced and interpreted (subtraction being one, but only one, such process). They place functional images at an epistemically significant remove from the neuronal activities they are used to study by generating disparities between displayed and real BIs. At the same time physiological and technological factors conspire to remove displayed BIs a step further from the neuronal activities. These factors raise questions that do not depend on the entanglements between FI and cognitive psychology emphasized by Poeppel, van Orden, Paap, Stufflebeam, and Bechtel.

Some Measurement Troubles

The connection between the records of radio and radiation signals from which functional images are produced and the neuronal activity the images are used to study is impressively indirect. One way to appreciate this is to compare image production to the use of a mercury thermometer to measure temperature. When a thermometer in good working order is properly applied under standard conditions to an object of uniform temperature, not too much below the freezing point of water or above the boiling point, the height of the mercury column depends on and varies directly with the very quantity it is used to measure. The thermometer will have been calibrated—in accordance with well-understood physical mechanisms through which the temperature influences the height of the column—to assign numbers in a principled way to temperatures. The technology by which graduated lines are placed on the thermometer introduces no unmanageable sources of significant error. The accuracy of the thermometer can be checked and adjusted using a variety of well-established methods. Many significant features of any given individual thermometer—length and choice of scale, for example—are matters of convention. Because readings from well-made thermometers produced according to different conventions are intertranslatable, and because the calculations required to translate and check the readings for consistency can be quite uncontroversial, the conventions of thermometer design, choice of scale, and so forth can be shown to be epistemically benign. The reliability of the best procedures for producing data by visually inspecting the thermometer and recording the relevant numbers is easily enough established to satisfy anyone but a general skeptic.

Functional images are another story. In most cases a positron emitted by a radioactive oxygen or fluorine atom travels a short distance before it encounters and interacts with an electron to produce a pair of gamma rays that travel away from one another along a straight line. Interactions that occur when one or both of them encounter a detector during a PET scan are entered into a signal detection record, which is then processed to determine—from such factors as the places and times at which the recorded detection interactions occurred—which of the detected gamma ray pairs are most likely to have originated from the same positron-electron interactions, and where in the brain those interactions occurred (Corbetta 1998, 96). Further computations are needed to determine, pixel by pixel, which BI colors should be displayed and where they should go. fMRI imaging takes advantage of the fact that the brief application of a magnetic field causes oxygen protons to precess coherently enough around a plane within the brain to produce a detectable radio signal, which lasts for a short time until the precessions become too random to continue to produce the signal. Because signal strengths and durations vary with oxygen levels, relative amounts of oxygenated hemoglobin in different brain regions can be computed from records of the signals the scanning equipment detects (Haxby et al. 1998, 123–24; note 4 above). Although fMRI signal detection records are considerably less noisy than their PET counterparts, they share a crucial feature: neither record tells us anything about BI levels or locations until heroic mathematical measures have been taken to decide, pixel by pixel, which BI magnitudes to indicate where.

Thus a functional image is by no means a record of visible effects produced by, and varying directly with, the BIs, let alone the neuronal activity of interest. The image is not to be confused with the signal detection record from which the pixel colors are computed. It is instead the result of extensive analysis and interpretation of the signal detection record. The equipment used to produce the picture is designed and employed in accordance with background beliefs about the neuronal quantities to be measured, the mechanisms that connect them to the BIs and that connect the BIs to the radioactivity or radio signals, and the mechanisms that allow for recording and detecting the signals.[7] The computations embody a host of statistical and other mathematical assumptions and conventions. The visual display is heavily constrained by all of this, in addition to practices and conventions involved in the design, choice, and use of software for the computations and in almost every other stage of image production. Whether or not the assumptions, practices, and conventions are epistemically innocent, they and the roles they play are far too different from their counterparts in the production of thermometer data to allow their innocence to be evaluated in anything like the same way.

Real versus Displayed BI Magnitudes

Individual PET images are so noisy and their error ranges are so wide that in one typical run "differences between pixel values of two scans of the same subject performing the same sensorimotor control task have a distribution with a standard deviation of 15–16%" (Haxby et al. 1991, 542).

Some error sources are physiological: the levels of BIs to which the equipment is sensitive may fluctuate in ways that have no systematic connection to task-related neuronal activity. Gamma rays from sources extraneous to the BIs, and coincidental spatial and temporal relations between detected gamma rays, contribute further errors. A great deal of noise comes from the workings of the scanning equipment (Haxby et al. 1991, 542). A typical strategy to correct for all of this is to perform averaging and other statistical operations on the results of BI estimates calculated from a number of individual scans. Unfortunately several features of PET technology conspire to favor the averaging of images from different subjects for this purpose. Ethical considerations argue against introducing as much radioactive material into a single subject as often as would be needed to reduce noise to a satisfactory level by intrasubject averaging (Steinmetz and Seitz 1991, 1158).

Furthermore, it takes ten minutes after ^{15}O ingestion for radiation levels to return to baseline, and it takes forty to sixty seconds of scanning to record enough signals to produce a useable PET image. This makes it impossible to make images at short enough intervals to avoid further error due to sources that operate between scans. Accordingly most investigators rely on intersubject averaging for error reduction.[8] The averaging washes out details of individual scans in such a way as to bias PET images toward high levels of BIs and toward "functional areas of either large extent [or] little inter-subject variability" (Steinmetz and Seitz 1991, 1158).

Here is one way this can happen. Suppose, as Steinmetz and Seitz suggest, that a given mental function is carried out through the neuronal activity of several small anatomical modules located at short distances from one another within a somewhat larger area of the brain. If their relative positions vary sufficiently from brain to brain,[9] averaging will obliterate signs of the geographical distribution of the relevant BIs. If their BI levels are not extraordinarily high, averaged images are likely to represent the area that contains them as relatively inactive (Steinmetz and Seitz 1991, 1154).

fMRI avoids some of these problems. It is relatively noise free, and it allows for the production of a great many images of the same subject from signals detected over shorter periods of time (Bavelier et al. 1997, 680–81). Thus intersubject averaging is not required for noise reduction, and subtraction images can be made from a single subject. But even so, intrasubject averaging is required to correct for BI magnitudes that are idiosyncratic to one or a very few performances of the same task. And intersubject averaging is required, for example, to correct for the anatomical and physiological idiosyncrasies of individual brains "for the purpose of comparison with previous data from [neurocognitive studies of the performances of subjects with brain] lesions and from [other] imaging studies" (Bavelier et al. 1997, 671).

Here, as with PET, information about levels and locations of BIs will be washed out. The same holds true, of course, for PET and fMRI subtraction.

Further biasing toward high BI levels and large anatomical regions results from threshold setting. Resting neurons fire at rates that are random with regard to task performance. Neurons involved in one task fire at rates that are random with regard to the performance of others. Some variations in blood and oxygenated hemoglobin levels are random relative to neuronal activities involved in mental functions. Accordingly, pixel-by-pixel probability thresholds are set in hopes of screening out signals from such factors. This biases PET and fMRI images further in favor of large increases of regional cerebral blood flow over baseline levels. Further mathematical manipulation is required for pixel registration. For example, image smoothing is required to deal with BI levels in regions that fall between the pixels of images used in intersubject averaging or subtraction (see, e.g., Watson et al. 1993, 81). The biological assumptions involved in setting threshold levels and the mathematical assumptions involved in pixel adjustment conspire to produce further disparities between actual and imaged levels and locations of BI activity. These are some of the sources of disparities between actual and visually displayed BI levels.

Some Anatomical Mapping Complications

PET and fMRI images are functional rather than structural because the equipment is set to respond to signals that vary with biological indicators of neuronal activity. But thus set, the equipment cannot (as can conventional MRI, CAT scan, and X-ray detectors) pick up signals indicative of anatomical landmarks. How then can colors indicating BI levels be arranged to indicate where in the brain these levels are to be found? Fortunately the anterior and posterior commissures and the upper, lower, anterior, and posterior edges of the brain exhibit distinctive and relatively stable BI profiles which enable the investigator to orient functional images to landmarks in MRI anatomical images of the subject's brain. Even so, precise PET or fMRI-MRI superimposition is difficult (Haxby et al. 1991, 542). Because of the noisiness of PET images, their superimposition onto MRI images requires considerable analysis. Statistical analysis is required to a significant (albeit lesser) extent for fMRI-MRI superimposition because fMRI signals from anatomical landmarks can be contaminated with signals indicative of BI levels in nearby tissue. Other methods are available for anatomical mapping, including projection onto a standard brain atlas instead of an MRI image of the subject's brain, but they too involve extensive computations.[10]

Further mapping problems arise because claims about the functionally significant anatomical divisions of the brain cannot be satisfactorily generalized if they are too closely tied to the geographical idiosyncrasies of one or only a few brains. Even normal brains differ to an observable extent from one another with regard to the shapes, sizes, and relative positions of gyri, sulci, and other anatomical landmarks. Thus in typical FI experiments each image of a "unique individual brain is transformed anatomically ('normalized') in such a way that its conformation matches that of a standard brain. A topographic map developed for the standard brain then serves as the map for the transformed brain" (Caviness et al. 1996, 566).

Instead of depicting a real brain, the standard map is a highly idealized and simplified picture designed to eliminate anatomical idiosyncrasies of individual brains.[11] Thus the shapes, sizes, and relative distances between anatomical landmarks displayed in the standard map will differ in various ways and to varying degrees from those of real brains. The question of how such idealization affects the reliability of functional images as indicators of the anatomical locations of BIs is complicated by peculiarities of the "locations" mentioned in the very questions that functional images are often used to investigate. The bearing of functional images on familiar claims about the involvement of Broca's and Wernicke's regions in speech will serve to illustrate this point.

In comparison to such structures as the optic nerve and the brain itself, Wernicke's and (to a considerably lesser extent) Broca's regions are anatomically vague. Furthermore, their locations are specified in significantly different ways. Whereas Broca's region is anatomically defined, the definition of Wernicke's region is largely functional.

Many anatomically defined brain regions have more or less vague boundaries. The most finely grained anatomical definitions follow

Brodmann's strategy of distinguishing one cortical area from another in terms of their cell types (see Carpenter and Sutin 1983, 653). For some regions (e.g., at the border of the striate cortex), the boundaries between cells of different types are "hair sharp." Other boundaries, like the border of the acoustic core area, are reasonably sharp, though less so. Other regions have "hazy boundaries which show within narrow limits a gradual transition" from cells characteristic of one area to cells characteristic of the neighboring area. Braak (1980, 112) maintains that, for anatomically defined regions of the brain, the "construction [by anatomical theorists] of the individual borders appears . . . as manifold as the construction of the areas themselves." Broca's area is defined by reference to anatomical landmarks even though it is not separated from the rest of the brain by a spatial gap, by surrounding membranes, or by sharp cytoarchitectural boundaries. Typical definitions identify it with "the foot, that is, the posterior third of the inferior frontal gyrus" (Bogen and Bogen 1976, 835) or with "portions of the pars triangularis and opercularis of the inferior frontal gyrus" in the dominant hemisphere (Carpenter and Sutin 1983, 703).

Wernicke's area lacks sharp anatomical boundaries for an entirely different reason. It is defined not just by reference to anatomical landmarks, but as a region posterior to the Rolandic fissure near the end of the Sylvian fissure, where lesioning or electrical stimulation is likely to produce certain types of speech disturbances. Even allowing for the fact that some theorists base their definitions on somewhat different disturbances, lesioning and stimulation produce the relevant disturbances at observably different locations in different brains (Bogen and Bogen 1976, 835–42).[12] Thus even though they mention anatomical landmarks, the standard characterization of Wernicke's area is essentially functional. Wernicke's is not the only cortical region defined in this way.[13] Although a functional and an anatomical definition may identify roughly the same anatomical region, they may not; one functionally defined area may correspond to different anatomically defined areas in different brains, whereas an anatomically defined area may be associated with different functions in different brains.

This situation introduces a new wrinkle into the old problem of understanding how more or less abstract and unrealistic idealizations apply to concrete, real-world systems. With regard to specific instances of Q1 or Q2, that problem is complicated by further questions of whether the vagueness of anatomical boundaries and the geographical differences between brain regions that satisfy one and the same functional definition significantly amplify or mitigate the errors generated by anatomical disparities between real and imaged brains.

Troubles with BI–Neuronal Activity Correlations

The interpretation of functional images is further complicated by striking differences between the connections of real and imaged BIs to functionally significant neuronal activity. Here are some of the complications.

Threshold Setting Filters Out Signs of Low-Level Neuronal Activity That May Be Cognitively Significant

In addition to statistical thresholds established to screen out spurious indications of above-threshold BI levels, thresholds are set in hopes of screening out levels that are too low to indicate functionally significant neuronal activity. Because there is no agreement on just what a genuine, but relatively low, change in blood flow (e.g., a change of less than 10 percent) indicates with regard to neural processing, thresholds are by convention set to screen them out. As a result, it is plausible that FI tells us about only "the most robust aspects of neural processing in any given task" (Gazzaniga et al. 1998, 110–11).

Thus FI contrasts with electrophysiological techniques that register weaker and more highly distributed neuronal activity but do not reveal nearly as much about its anatomical location.

Problems of Timing

FI's temporal resolution is coarse. It takes forty to sixty seconds for a PET scanner to detect enough signals for a comparison of blood levels at a distance of ten to fifteen millimeters (Corbetta 1998, 98).[14] fMRI requires only a few seconds for detections required to compute significant differences in oxygenated hemoglobin levels a few millimeters apart (Haxby et al. 1998, 130). But even so, a complex mental task may be performed in less than a second, and one of its components may require less than a hundred milliseconds (Posner and Raichle 1994, 143). Thus displayed BI levels may be associated with neuronal activity involved in the simultaneous and sequential performance of a number of different mental functions accomplished during the time it takes to collect enough signals for a single image.

A second timing problem arises from the fact that functionally significant neuronal activities and the associated BIs are unlikely to develop at the same rates. BI levels rise and fall gradually during an interval near the time during which the brain carries out the associated mental function. Although the equipment cannot hope to detect BI changes that occur during the time required to produce a single image, the comparison of different images may provide a rough indication of its development. But in order to interpret functional images to check claims involving neuronal activity, the investigator must choose a model for its development. Neuronal activity is commonly modeled as rising immediately to a level above baseline, remaining there for a time, and then dropping immediately back to baseline. According to Haxby this convention is adopted for reasons of computational convenience rather than theoretical or empirical support (Haxby et al. 1998, 126). In fact, although it seems appropriate for the development of a burst of action potentials in a single cell, it probably fails to fit the overall activity of groups of cells whose cooperation is required for the mental functions FI is used to study.[15]

Inhibition Troubles

FI is not sensitive to differences between inhibitory and excitative activity (Corbetta 1998, 98). This may account for a surprising result Sidtis et al. obtained by studying PET images for the following conditions:

1. *Baseline condition.* Blindfolded subjects in a dimly lit room whose auditory canals were plugged with silent earphones were "required to remain awake and quiet."
2. *Syllable repetition.* Subjects repeated the syllables *pa, ta, ka* as fast as they could.
3. *Sustained phonation.* Subjects pronounced *ah* for extended periods without consonants.
4. *Repetitive lip closure.* Subjects silently moved their lips as if they were pronouncing *pa* over and over as fast as they could (Sidtis et al. forthcoming).

Syllable repetition is a complex task requiring lip closure and the production of *ah* sounds. Thus the condition 3 and 4 tasks approximate to some extent components of syllable repetition, and one would expect higher BI levels for condition 2 than for either of them. If image subtraction discriminates among BI levels associated with different components of a complex task, task-minus-baseline subtraction images should display lower BI levels for conditions 3 and 4 than for condition 2. But visual inspection of the simple subtraction images "suggested that [neuronal] 'activation' was generally greater in subcortical areas during phonation and lip closure compared to syllable production" (Sidtis et al. forthcoming).

BI levels were "significantly greater" for lip closure than for syllable repetition in the left and right putamen, the left and right caudate, and the left and right thalamus (but not in the left and right superior temporal gyri and the left and right transverse temporal gyri). BI levels were significantly higher for phonation than for syllable repetition "in the left and possibly the right putamen" (but not in the right superior temporal gyrus and the left and right transverse temporal gyri) (Sidtis et al. forthcoming).[16] Assuming (as FI must) that BI levels are roughly proportional to levels of neuronal activity, a possible explanation for this finding would be that the excess neuronal activity indicated in condition 3 serves to inhibit lip closure, which would interfere with phonation, and, in condition 4, with production of the vowel sound during silent lip closure.[17] Since no such inhibitory activity is required for syllable repetition, this could explain why lower BI levels are displayed for condition 2 than for conditions 3 and 4. Alternative explanations are possible, but this does not affect the moral I want to draw from this example.[18] Displayed BI levels will not tell us whether inhibition is the right explanation for the results reported by Sidtis et al. Lacking as it is in marks by which BIs associated with excitatory and inhibitory neuronal activity can be distinguished from one another, it need make no difference to the color display for condition 4 whether the caudate, the putamen, or another region excites motor neurons to move the lips or inhibits neurons that would otherwise produce the vowel sound (and similarly for condition 3).[19] This is an instance of a general epistemological problem. Q1 and Q2 require the investigator to differentiate brain regions whose neuronal activity belongs to the execution of a target mental function from areas whose activities serve other purposes. The general problem is to understand how FI can help with this. I believe the point that these results illustrate is that FI's inability to mark excitatory as opposed to inhibitory BIs is yet another instance of the general problem.

To What Genre Do Functional Images Belong?

Peter Galison contrasts the use of "picturing machines" to produce images of individual effects with the use of "electronic counters coupled with electronic logic circuits [to] aggregate masses of data [from a number of individual occurrences in order] to make statistical arguments for the existence of a particle or effect" (Galison 1997, 19). Bubble chamber experiments exemplify the former. Here things are arranged so that interactions involving charged particles produce tracks in a confined space where they can be photographed by a camera set to shoot pictures at predetermined intervals. The visual similarities between the resulting images and the tracks they represent are sufficient to enable an investigator to tell by looking whether individual photographs are indicative of occurrences of particle interactions of interest. By contrast counters and logic circuits are not used to produce pictures, let alone pictures of single items. Here details of individual occurrences are blurred, lost, or replaced in the course of the statistical manipulations of data. Unlike the statistical approach, which "consciously sacrifices the detail of the one for the stability of the many," images (e.g., photographs or naturalistic drawings of individual items) can preserve details obscured by statistical analysis at the risk of directing the investigator's attention to uninformative flukes and oddities (Galison 1997, 20).

Functional images belong to a hybrid genre. It is crucial to their use as evidence that they look somewhat like naturalistic pictures of real brains. But instead of real brains, they depict more or less unrealistic models from which some features peculiar to individual brains have been abstracted away and in which others have been replaced by idealizations. Statistical manipulations are essential to their production— so much so that they are sometimes explicitly labeled "statistical parametric maps" (see, e.g., Frith et al. 1991, 1140, 1143). I am indebted to Florence Fogelin for suggesting in conversation that in this respect functional images resemble tables, graphs, and charts, whose visual layouts are designed to help their users grasp relationships between quantities calculated from raw data. But at the same time they exploit features of the visual appearance of the brain for this purpose in something like the way some geographical maps use symbols and coded colors to indicate regional differences in population density, agricultural production, and other statistically calculated quantities. Thus even though FI sacrifices anatomical details to reduce noise and to isolate BIs associated with particular subtasks, it still exploits visual appearances in ways that place its images somewhere between detailed naturalistic pictures and graphs or charts.

Unlike data, functional images are tapestries woven on a warp of assumptions and background beliefs from fibers from biology, psychology, physics, chemistry, statistics, computer science, and a variety of technologies that figure in different ways in the production of FI equipment and in the design and execution of the experiments that produce functional images. There are in addition conventions, practices, and seat-of-the-pants decisions, some of which are idiosyncratic to a particular research program, an individual laboratory, or even a single investigator. Some of these vary from laboratory to laboratory, and even from time to time in the same laboratory.

Some Epistemological Questions

Stufflebeam and Bechtel (1997, S96) are certainly right to emphasize the value of FI as a research tool. But they also think that as measures of "the brain activity associated with particular cognitive tasks [functional images are] by now relatively unproblematic." I disagree—not because of the commitments to cognitive models discussed earlier, but because of the peculiarities of the FI genre just noted, and because of the distinctions between displayed BIs and functionally significant neuronal activities. The problems they pose have to do with information loss[20] in imaging production and with questionable assumptions that arise outside cognitive psychology.

Information Loss

FI employs high-tech versions of strategies that Claude Bernard warned physiologists against. According to the methodology he advocated, conclusions about physiological phenomena are to be arrived at by detailed chemical analyses of bodily substances and wastes, and from precise measurement of the fluid, air, and food intakes; temperature; pulse; and other physiological parameters of individuals. Bernard argues that in many cases these quantities vary so greatly from individual to individual, and in the same individual under different conditions, that a description of the average case "will never be perfectly matched in nature." Except in atypical cases involving quantities that vary only slightly, statistical averaging hinders the physiologist by obscuring crucial details to such an extent that "the true relations of phenomena disappear in the average." As an illustration, he recounts the story of a misguided physiologist who analyzed urine from a railroad station urinal used by travelers from all over Europe in order to determine the chemical composition of average European urine (Bernard 1957, 125, 134–35). Since no individual produces such urine, there is no reason why the details of its chemical makeup should have anything interesting to tell us about human physiology. In the sense in which average European urine is not real urine, and the magnitudes of its chemical quantities are not real quantities, the BIs displayed in functional images are not real BIs, and their displayed locations are not real locations. I think FI has an impressive track record despite this kind of information loss. The question is how to explain this.

Questionable Assumptions

Investigators cannot produce or draw inferences from functional images without committing themselves—implicitly at least—to false or dubious assumptions. Problematic as it is, this is not a matter of Hansonian, Kuhnian, Feyerabendian, or Churchlandish[21] theory loading (Hanson 1958, 4–30; Kuhn 1970, 62–63, 52–65, 111–35; Feyerabend 1985, 17–36; Churchland 1992, 5–30). Producing, perceiving, and drawing conclusions from functional images need not, and typically does not, commit the investigator to the very claims the images are used to test, or to claims that imply them or their denials, or to ideas that prejudge the very issues under consideration. That is why functional images used to test Petersen's hypothesis (h) can be seen, and described, in the same way by friends and foes of (h) (and of the Baddely supposition that (h) bears on). Choosing software that uses certain algorithms to smooth, average, or subtract images need not, and usually is not, the same thing as designing an FI experiment whose results cannot be perceived or described as counting against one's own, or a competing, hypothesis about the location of functionally significant neuronal activity. By biasing an experiment against low BI levels in small regions, threshold settings can rule out experimental outcomes that would count in favor of (or against) some hypotheses, but not *every* hypothesis, in which an investigator might be interested. Furthermore, threshold choice is usually determined by caution in the face of uncertainty over the significance of relatively small BI levels, not by the investigator's commitment to the very claims an FI experiment is used to investigate.

Thus to acknowledge the obvious and substantial influence of theoretical and practical commitments on the choice, design, and conduct of experiments and the interpretation of their results is by no means to say that functional images are theory laden in the sense suggested by Kuhn et al. But neutral as they are with regard to the claims the images are used to evaluate, some assumptions involved in image production and interpretation are unconfirmed, if not dubious. For example, threshold setting conventions implicitly commit the investigator to the insignificance of BI levels less than 10 percent above baseline. Others, like the assumption that functionally significant neuronal activity develops in accordance with a square wave function, are probably false. I believe FI's track record shows that its results are better than the assumptions it involves.[22] The question raised by its commitments to false or dubious assumptions is how this is to be explained.

What should we make of the information loss and the involvement of questionable assumptions in the production of functional images? On the one hand, they make it unreasonable to expect any impressive degree of accuracy and precision from PET and fMRI imaging, at least in their present state. Under reasonably good conditions functional images can indicate relative levels and locations of BIs for sufficiently large differences and sufficiently large and well-separated anatomical regions. Under such conditions it is also reasonable to expect the images to provide rough indications of locations and relative levels of neuronal activity. But it is hard to see how they could provide much beyond coarse-grained approximations. There are qualitative claims about neuronal activity (associated with sufficiently high BI levels) occurring at roughly the same time as the performance of a mental task that can be justified by appeal to FI evidence. But the separations discussed previously between displayed BIs and functionally significant neuronal activity make it hard to see how functional images can show us just which part of the brain it is whose neuronal activity is peculiarly associated with a given task (Q1) let alone just what contribution that activity makes to its performance (Q2).[23] On the other hand, investigators using FI to study a variety of different questions have obtained good enough results to make it quite implausible that functional image evidence is by nature epistemically defective.

How much epistemic harm is actually done by the features of FI that I have been emphasizing depends upon the question under investigation and the strategy used to look for an answer. There are cases in which functional images can contribute to the justification of interesting claims about mental and brain functions despite information loss and the questionable practical and theoretical assumptions involved in their production. These are cases in which the investigator's purposes can be served by coarse-grained, qualitative findings about neuronal activity in imprecisely defined regions of the brain. For example, Passingham (1993, 13–15) was able to use PET images (along with evidence of other kinds) to justify the claim that what he calls the motor and premotor areas of the macaque brain are for certain purposes acceptably good models of the corresponding areas of the human brain. These areas of the frontal lobe are adjacent to one another, hazily bounded, and considerably larger than the anatomical modules Steinmetz and Seitz believed were hidden by image averaging. The results of lesioning, single-cell recording, and other experimental techniques suggested that the macaque motor area is much more involved in learned or voluntary behavior than the macaque premotor area, and that the premotor area is peculiarly connected to unlearned, involuntary behavior (Passingham 1993, 22). The experimental techniques used on the macaques are too invasive and otherwise inappropriate for application to healthy human subjects. In their place Passingham used PET to image normal human subjects as they repeated such voluntary movements as raising an arm by flexing a shoulder, opening and closing a hand, touching a forefinger to the other fingers in turn, and abducting an index finger. The displays Passingham obtained by subtracting resting from task performance images indicated significant increases of neuronal activity in the motor area during the performance of these tasks. Assuming that the relevant human motions are voluntary in roughly the same sense as voluntary macaque movements, these findings support the adequacy of the macaque model (Passingham 1993, 33–35).[24]

Functional images could serve this purpose because—despite their imprecision and proneness to error—the subtraction images displayed unmistakably high BI levels during task performance, and these displays could not plausibly be discounted as artifacts of image production. Despite unrealistic anatomical idealizations, anatomical vagueness, and the slippages between BI levels and functionally significant neuronal activity, the displayed BIs indicated impressively high levels of neuronal activity well within the motor area, and impressively low levels within the premotor area. For Passingham's purposes, this evidence survives the Bernardian objection with its epistemic credentials intact, despite the details it obscures. And even if the most questionable assumptions involved in their production were all false, it is implausible that they could be false for reasons that would render the images unreliable as evidence for the adequacy of the macaque model. I think something like the same thing holds for the Petersen's STG experiments.

This makes it natural to ask how functional images can contribute to the investigation of other neuroscientific questions, and whether there is any informative, systematic story to tell about the reliability of the research strategies that employ them. That is an issue I hope empiricists will pursue—along with the others I have tried to raise in this chapter.

NOTES

I am indebted to Joseph E. Bogen for help of various kinds and to J. J. Sidtis for generously providing and answering questions about preprints of some of his recent work.

1. This is Jim Woodward's and my surrogate for more traditional distinctions, for example, between perceptual evidence, observational evidence, and observation reports and conclusions drawn from them (see Bogen and Woodward 1988, 1992). Kuhn, Feyerabend, Hanson, and Churchland blur this distinction in ways that disqualify them as empiricists. I defer to Sellars scholars as to whether or not he qualifies.

2. For more details see "Tutorial: Nuclear Physics and Tomography" on the Web at http://www.crump.ucla.edu/lpp/nuclearphysics/imagerecon.html.

3. For more details see M. S. Cohen and S. Y. Bookheimer, "Functional Magnetic Resonance Imaging," on the Web at http://fmri.ucsd.edu/~stark/fmri/FMRI-TINS.html, and M. S. Cohen, "Basic Magnetization Physics," on the Web at http://fmri.ucsd.edu/~stark/basicmr/basicMRphcs.htm.

4. For example, a subject sitting in a dark room, wearing earplugs to block auditory stimuli, may be instructed to relax without thinking about, or doing, anything in particular.

5. Petersen supposed the neuronal activity should have something to do with "semantic processing or association between words" (Petersen and Fiez 1993, 525).

6. These results are of interest because they tend to rule out certain possibilities, including that the BI levels peculiar to the rhyming task could have been

due to activities that do not involve phonological encoding, such as "whatever processes occur automatically in word recognition, . . . recollection/reactivation of the task demands [which would be required for the performance of any task and would therefore have no special connection to the rhyming task] . . . [or] selection/ generation of the response" (Démonet et al. 1996, 371).

7. For the relevant notion of mechanisms, see Machamer et al. (2000) and Craver and Darden (this volume).

8. In PET averaging, groups of six to eight different subjects are considered small (Watson et al. 1993, 91). For an exception to the practice, see Watson et al. (1993, 81).

9. There is enough evidence from stimulation and other experiments that do not use functional images to make all of this quite plausible (see, e.g., Steinmetz and Seitz 1991, 1151, 1154).

10. For lack of better terminology, I have used and will continue to use the term *functional image* in connection with individual PET or fMRI images as well as for the results of functional-structural superimpositions, subtraction images, and so on. I rely on context for disambiguation. The same goes for the terms *fMRI* and *PET image*.

11. For the most commonly used standard map see Tailairach and Tournoux (1988). For comments on it see Caviness et al. (1996).

12. Bogen and Bogen report a bewildering variety of geographically different locations identified as "Wernicke's area" by accepted authorities, including Geschwind, Head, Marie, Penfield, and Wernicke himself. Some authorities (including Wernicke!) "locate" Wernicke's area by placing a small mark inside a good-sized region on a brain map without specifying any boundaries for it within that region (Bogen and Bogen 1976, 835). In view of this it is surprising how often Wernicke's area is referred to as if it were anatomically defined (see, e.g., Zatorre et al. 1996, 28).

13. For example, Watson et al. identified visual area V5 by "determining the region of maximal rCBF change [in] . . . a region of the occipital lobe in which cortical activation by visual motion might be expected in each individual" (Watson et al. 1993, 83). They found that the geographical location V5, so defined, differed observably from brain to brain. Most landmarks by reference to which brain regions are located in MRI images of individual brains and in standard atlases are fissures or furrows and bulges of neuronal tissue that bound the regions. Their precise locations depend upon the way cortical tissue folds as it develops, and therefore they vary from brain to brain. Furthermore, even if the brains of individuals could be flattened out to correct for these variations, the spatial locations of at least some functionally defined areas in different brains would still differ observably.

14. According to Corbetta (1998, 98), "smaller differences in location (5–7 mm) can be resolved when single foci of activity are compared across different subtraction images," but the computations required for the production and comparison of subtraction images must be paid for by loss of information.

15. I am indebted to Newton Copp (personal communication) for explaining that the assumption seems more plausible for certain central motor activities in

which coordination involves a great deal of mutual inhibition than for other functions. When a circuit controlling one motion becomes active, it suppresses the activity of circuits whose activity would produce contradictory movements, so that its activity in relation to that of the competing circuits is suddenly and greatly amplified. It is generally considered to be different with perceptual and cognitive systems, in which "most neurons have a spontaneous firing rate when unstimulated such that information can be coded in graded increases . . . and decreases in activity." These are "graded changes in the balance of firing among neurons [or] groups of neurons, not discrete 'quantal' changes in firing." I am indebted to Jim Groome for emphasizing that the activity of a number of neurons may depend partly on the overall frequency of action potential spiking; a number of action potentials spaced 10 milliseconds apart produce considerably more neurotransmitter than exactly the same number of action potentials spaced 200 milliseconds apart.

16. It may be objected that lip closure and phonation are not really components of syllable repetition. I believe that differences between lip closure and phonation and components of syllable repetition cannot account for the results reported by Sidtis. He acknowledges that the lip closure required in condition 4 "is not the exact articulatory gesture made for *ta* and *ka*." But even so, since the production of the *ah* sound and the lip closure required for the first syllable are among the things condition 2 requires the subject to do, the use of subraction methodology to isolate BI levels associated with different components of a complex task from one another seems to commit the investigator to the assumption that BI levels will be lower for lip closure than for syllable repetition. Phonation (sustaining the *ah* sound) appears to be somewhat different from the production of the short, unsustained *ah* sound required for syllable repetition. Sidtis does not address this point explicitly, but it still seems plausible that, according to the assumptions underlying the use of image subtraction, BI levels for condition 3 should be lower than those for condition 2.

17. This hypothesis has some plausibility because the caudate, whose BI levels appear to be elevated in conditions 3 and 4, typically plays an inhibitory role.

18. For example it could be that lip closure and phonation require more attention and supervision than syllable repetition, perhaps because the latter task is relatively easier or more familiar.

19. Of course it is easy to think of cases in which some inhibitory activity is involved in the target task itself. For example, a considerable amount of inhibition is required for the successful execution of fine motor movements. But this does not make the inability of functional images to distinguish inhibitory from excitatory activity any less troublesome. If a functional image is to exhibit signs of neuronal activity associated with the target task (as required for Q1), it would have to distinguish them from signs of neuronal activity whose sole function is to inhibit activity in neurons that make no contribution to the performance of the target task.

20. I have no theory of information; throughout this chapter the term *information* is used loosely and colloquially.

21. For objections to Churchland's and related versions of the idea that empirical evidence is theory laden, see Bogen and Woodward (1992).

22. Duhem's discussion of Newton's arguments for universal gravitation, and Laymon's discussions of related cases show that this problem is by no means peculiar to neuroscience, or biology, let alone to FI (Duhem 1982, 190–95; Laymon 1988).

23. In this connection it is important to bear in mind, as Stufflebeam and Bechtel (1997, S105–S106) emphasize, that supplementation by evidence and background knowledge from other quarters is essential to the reliable interpretation of FI evidence. But it will take more than an understanding of the supplementations to explain the epistemic virtues of functional images, and how they possess them in view of the epistemic peculiarities I have been sketching.

24. Needless to say, Passingham did not rest his case for the adequacy of the macaque model on the results of this experiment. He used it to complement other evidence from FI and from a variety of other methodologies. Its importance lies in the fact that, unlike the animal and some of the other human evidence, the PET images are derived from anatomically and functionally normal human brains.

REFERENCES

Baddeley, A. 1998. *Human Memory: Theory and Practice*. Boston: Allyn and Bacon.

Bavelier, D., D. Corina, P. Jezzard, S. Radmanabkan, V. P. Clark and A. Karni, A. Prinster, A. Braun, A. Larwani, J. P. Rauscheker, B. Turner, and M. Neville. 1997. "Sentence Reading: A Functional MRI Study at 4 Tesla." *Journal of Cognitive Neuroscience* 9(5): 664–86.

Bernard, C. 1957. *An Introduction to the Study of Experimental Medicine*. New York: Dover.

Bogen, G., and J. E. Bogen. 1976. "Wernicke's Region: Where Is It?" *Annals of the New York Academy of Sciences* 280: 834–43.

Bogen, J. E. 1976. "Hughlings Jackson's Heterogram." In D. O. Walter, L. Rogers, and J. M. Finzi-Frii, eds., *BIS Report #42*. Los Angeles: University of California at Los Angeles, 148–50.

Bogen, J., and J. Woodward. 1988. "Saving the Phenomena." *Philosophical Review* 97: 303–57.

———. 1992. "Observations, Theories, and the Evolution of the Human Spirit." *Philosophy of Science* 59(4): 590–611.

Braak, H. 1980. *Architectonics of the Human Telencephalic Cortex*. New York: Springer-Verlag.

Carpenter, M. B., and J. Sutin. 1983. *Human Neuroanatomy*, 8th ed. Baltimore: Williams and Wilkins.

Caviness, V. S., Jr., J. Meyer, N. Makkis, and D. Kennedy. 1996. "MRI-Based Topographic Parcellation of Human Neorcortex: An Anatomically Specified

Method with Estimate of Reliability." *Journal of Cognitive Neuroscience* 8(6): 566–87.

Churchland, P. 1992. *A Neurocomputational Perspective.* Cambridge, Mass.: MIT Press.

Corbetta, M. 1998. "Functional Anatomy of Visual Attention in the Human Brain." In R. Parasuraman, ed., *The Attentive Brain.* Cambridge, Mass.: MIT Press, 95–122.

Démonet, J. E. F., J. A. Fiez, S. E. Petersen, and R. Z. Zatorre. 1996. Reply. In Poeppel 1996, 352–79.

Duhem, P. 1982. *The Aim and Structure of Physical Science,* transl. P. Wiener. Princeton, N.J.: Princeton University Press.

Feyerabend, P. K. 1985. *Realism, Rationalism and Scientific Method.* Cambridge: Cambridge University Press.

Frith, C. D., K. J. Friston, P. F. Liddle, and R. S. J. Aackowian. 1991. "A PET Study of Word Finding." *Neuropsychologia* 29(12): 1137–48.

Galison, P. 1997. *Image and Logic.* Chicago: University of Chicago Press.

Gazzaniga, M. S., R. B. Ivry, and G. R. Mangun. 1998. *Cognitive Neuroscience: The Biology of the Mind.* New York: Norton.

Hanson, N. R. 1958. *Patterns of Discovery.* Cambridge: Cambridge University Press.

Haxby, J. V., C. L. Grody, L. G. Ungerleider, and B. Horowitz. 1991. "Mapping the Functional Neuroanatomy of the Intact Human Brain with Brain Work Imaging." *Neuropsychologia* 29(6): 539–55.

Haxby, J. V., S. M. Courtney, and V. P. Clark. 1998. "Functional Magnetic Resonance Imaging and the Study of Attention." In R. Parasuraman, ed., *The Attentive Brain.* Cambridge, Mass.: MIT Press, 123–42.

Kuhn, T. 1970. *The Structure of Scientific Revolutions.* Chicago: University of Chicago Press.

Laymon, R. 1988. "The Michelson-Morely Experiment and the Appraisal of Theories." In A. Donovan, L. Laudan, and R. Laudan, eds., *Scrutinizing Science.* Baltimore: Johns Hopkins University Press, 245–66.

Machamer, P., L. Darden, and C. F. Craver. 2000. "Thinking about Mechanisms." *Philosophy of Science* 67: 1–25.

Passingham, R. 1993. *The Frontal Lobes and Voluntary Action.* Oxford: Oxford University Press.

Petersen, S. E., and J. A. Fiez. 1993. "The Processing of Single Words Studied with Positron Emission Tomography." *Annual Review of Neuroscience* 16: 509–30.

Poeppel, D. 1996. "A Critical Review of PET Studies of Phonological Processing." *Brain and Language* 55: 317–51.

Posner, M. I., and M. E. Raichle. 1994. *Images of Mind.* New York: Scientific American Library.

Sidtis, J. J., S. C. Strother, J. R. Anderson, and D. A. Rottenberg. Forthcoming. "Are Brain Functions Really Additive?" *Neuroimage.*

Steinmetz, H., and R. J. Seitz. 1991. "Functional Anatomy of Language Processing: Neuroimaging and the Problem of Individual Variability." *Neuropsychologia* 29(12): 1149–61.

Stufflebeam, R. S., and W. Bechtel. 1997. "PET: Exploring the Myth and the Method." In L. Darden, ed., *Proceedings of the 1996 Bicentennial Meeting of the Philosophy of Science Association,* Part II, Supplement 64(4): S95–S106.

Talairach, J., and P. Tournoux. 1988. *Co-Planar Stereotaxic Atlas of the Human Brain.* New York: Thieme.

Van Orden, G., and K. R. Paap. 1997. "Functional Images Fail to Discover Pieces of Mind in the Parts of the Brain." In L. Darden, ed., *Proceedings of the 1996 Bicentennial Meeting of the Philosophy of Science Association,* Part II, Supplement 64(4): S58–S94.

Watson, J. D. G., A. Meyers, R. S. J. Frackiowak, J. V. Hajnal, R. P. Wood, J. C. Mazziota, S. Ship, and S. Zeki. 1993. "Area V5 of the Human Brain: Evidence from a Combined Study Using Positron Emission Tomography and Magnetic Resonance Imaging." *Cerebral Cortex* 3: 79–94.

Zatorre, R. J., E. Meyer, A. Giedde, and A. C. Evans. 1996. "PET Studies of Phonetic Processing of Speech: Review, Replication, and Reanalysis." *Cerebral Cortex* 6: 22–30.

10

Extrapolation from Animal Models

Social Life, Sex, and Super Models

Kenneth F. Schaffner
University Professor of Medical Humanities and Department of Philosophy,
George Washington University, Washington, D.C.

Much of our knowledge in the neurosciences is based on simple animal models. Examples in the history of the neurosciences abound and include the giant squid model so fundamental to Hodgkin and Huxley's model of action potentials, through to the work of Kandel and his colleagues on *Aplysia*. Much of contemporary biology including the neurosciences is increasingly focused on a small number of organism types. These have been termed "model organisms," and prominent leaders of the scientific community have defended the special attention paid to them. Bruce Alberts is a noted biologist, current president of the U.S. National Academy of Sciences, and the coauthor of a highly influential textbook, *The Molecular Biology of the Cell* (Alberts et al. 1994). He recently asked a rhetorical question about one of these model organisms, the common roundworm, *Caenorhabditis elegans:*

Why should one study a worm? This simple creature is one of several "model" organisms that together have provided tremendous insight into how all organisms are put together. It has become increasingly clear over the past two decades that knowledge from one organism, even one so simple as a worm, can provide tremendous power when connected with knowledge from other organisms. And because of the experimental accessibility of nematodes, knowledge about worms can come more quickly and cheaply than knowledge about higher organisms. . . . We can say with confidence that the fastest and most efficient way of acquiring an understanding of ourselves is to devote an enormous effort trying to understand these and other, relatively "simple" organisms. (1997, xii, xiv)

This efficiency is underscored when understanding the organism, if not ourselves, involves a *genetic* component. Many of the recent advances in neuroscience (though not all, as I point out later) rely on genetically based information. This holds not only for special strains of knockout mice (about which more subsequently), but also for many of our inquiries into mechanisms that are genetically based or genetically regulated.

The field that studies the role of genes' influences on behavior is known as behavioral genetics, and it is in this field that a number of neuroscientific advances have occurred. In June 1994, *Science* magazine devoted a special issue to the topic of "Genes and Behavior," including essays by a number of the major contributors to the field of behavioral genetics. In two overview essays by *Science*'s news staff, Mann and Barinaga stress that one of the main reasons that behavioral genetics has flourished in recent years is the "huge accumulation of data about hereditary influences in animal behavior" (Mann 1994, 1686); Barinaga (1994) entitled her article: "From Fruit Flies, Rats, Mice: Evidence of Genetic Influence." Essays by Hall (1994), Takahashi et al. (1994), and Thomas (1994) develop the details supporting these claims.

The importance of animal models in behavioral genetics as it relates to neuroscience is underscored in a chapter on "Genes and Behavior" in the textbook *Essentials of Neural Science and Behavior* by Kandel et al. (1995). This chapter, written by Greenspan et al., summarizes much of the recent research in comparative behavioral genetics. They write that "Extensive efforts are being made to understand the connections between genes and behavior in four organisms: the nematode *Caenorhabditis elegans*, the fruit fly *Drosophila melanogaster*, the laboratory mouse *Mus musculus*, and ourselves, *Homo sapiens*. Each has unique behavioral characteristics and each offers a different set of advantages and disadvantages for pursuing the study of genes and behavior" (1995, 556). In the remainder of this chapter, I discuss recent work involving these four organisms and the extrapolation issue, not only among the three model organisms on which I draw, but also to human beings.

First, the Worm

Caenorhabditis elegans is a tiny worm that has become the focus of a large number of research projects in many countries: projects that examine its genetics, development, nervous system, and behavior. In connection with the latter two areas, several groups of investigators (among them the laboratories of Avery, Bargmann, Chalfie, Horvitz, Lockery, Rankin, and Thomas) have worked to tie together the behavior of the organism and the underlying neural circuits and molecular processes implemented in those circuits. (Two other recent papers describe the early to middle years of worm research, essentially the period 1970–84; see de Chadarevian 1998; Ankeny 2000.) Here I concentrate on work done mostly in the 1990s, but I also provide some relevant background. Though behavior is quintessentially organismal—it is the organism as a whole that orients, moves, and mates—the explanations are devised at the molecular and neurocircuit levels and tested in populations using protocols that span many levels of aggregation. Following a quick review of the main relevant features of *C. elegans,* I describe some of these circuits and then examine two accounts that offer molecular and neurophysical details.

Some Basic Worm Facts

The adult hermaphrodite has 959 somatic nuclei and the male 1,031; there are about 2,000 germ cell nuclei (Hodgkin et al. 1995). The haploid genome contains 9.7×10^7 nucleotide pairs, organized into five autosomal and one sex chromosome (hermaphrodites are XX, males XO), comprising about 19,000 protein-coding genes (*C. elegans* Sequencing Consortium 1998). The organism can move itself forward and backward by undulatory movements and responds to touch and a number of both attractive and repulsive chemical stimuli. More complex behaviors include egg laying and mating between hermaphrodites and males (Wood 1988, 14). The nervous system is the largest organ, being composed, in the hermaphrodite, of 302 neurons, subdivided into 118 subclasses, along with 56 glial and associated support cells; there are 95 muscle cells on which the neurons can synapse. The neurons have been fully described in terms of their location and synaptic connections (Ankeny 2000). The neurons are essentially identical from one individual in a strain to another (Sulston et al. 1983; White et al. 1986) and form approximately 5,000 synapses, 600 gap junctions, and 2,000 neuromuscular junctions (White et al. 1986). The synapses are typically "highly reproducible" from one animal to another, but they are not identical.[1]

In 1988 Wood, echoing Brenner's earlier vision, wrote that "The simplicity of the C. *elegans* nervous system and the detail with which it has been described offer the opportunity to address fundamental questions of both function and development. With regard to function, it may be possible to correlate the entire behavioral repertoire with the known neuroanatomy" (1988, 14). Seemingly C. *elegans* is indeed what Robert Cook-Deegan (1994, 53) called "the reductionist's delight"—it also exemplifies what my colleague Horace Freeland Judson calls a "brute force" methodology. It has turned out, however, to be very difficult to tie the behavior to the neuroanatomy in any simple way.

As has been known for some time, this is partly because of the small size of the animal—it is difficult to study the electrophysiological or biochemical properties of individual neurons that are about two micrometers in diameter (Chalfie and White 1988, 338; Goodman et al. 1998). It has become standard practice to use the much larger neurons in another, closely related nematode, *Ascaris suum*, that permits some analogical inferences to be drawn about the neurons of C. *elegans*. Only very recently have patch clamping and intracellular recordings from C. *elegans* neurons begun to be feasible (Raizen and Avery 1994; Avery et al. 1995; Goodman et al. 1998). And it continues to be the case that information obtained from *Ascaris* plays an essential role in modeling neuronal interactions. In her 1993 review article Bargmann wrote that "heroic efforts" have resulted in the construction of a wiring diagram for C. *elegans* that has "aided in the interpretation of almost all C. *elegans* neurobiological experiments." But she went on to say that "neuronal functions cannot yet be predicted purely from the neuroanatomy. The electron micrographs do not indicate whether a synapse is excitatory, inhibitory, or modulatory. Nor do the morphologically defined synapses necessarily represent the complete set of physiologically relevant neuronal connections in this highly compact nervous system." She added that the neuroanatomy must be *integrated* with other information to determine "how neurons act together to generate coherent behaviors," for instance, with studies that utilize laser ablations of individual neurons, genetic analysis, pharmacology, and behavioral analysis (Bargmann 1993, 49–50).

This is still the case. Significant progress has been made, but the details of the circuits in the worm are still not directly observable. A combination of methods continues to be brought to bear on the neuro-physiology of *C. elegans,* and I describe these in the next section.

Worm Circuit Identification and Analysis

We may think of circuits as the first layer down from the organismal where the explanations of behavior begin. This neural circuit level of analysis has been pursued vigorously for about a dozen years in *C. elegans.* Some of the known circuits include those accounting for touch sensitivity, a tap withdrawal reflex, thermotaxis, and chemo-taxis. In this chapter I cannot present the details of each of these various circuits, but a description of the touch sensitivity circuit of Chalfie et al. in its simplified form can be found in Schaffner (1998a); this is a good illustration of the connectivity of neurons in the worm related to a set of behaviors. It involves a reflex circuit that generates a movement away from a fine touch stimulus—typically the stroking of the animal with a thin hair. This circuit receives input from five touch receptor neurons (ALML, ALMR, PLML, PLMR, and AVM); it then acts through five pairs of interneurons and motor neurons on muscle cells to generate forward and backward movements. A closely related (in fact, overlapping) circuit for a tap withdrawal reflex has been characterized in Rankin's laboratory (Wicks and Rankin 1995; Wicks et al. 1996). In addition, a fairly detailed neural network for ther-motaxis behavior in *C. elegans* was published by Mori and Ohshima (1995). Finally there is the chemotaxis circuit that Bargmann's labora-tory has investigated in a preliminary way. The chemotaxis circuit will be further refined using laser ablation tools in Lockery's laboratory. (A diagram of a simplified form of the chemotaxis circuit is provided as figure 10.1.)

All of these circuits are at present strongly underdetermined by any *direct* evidence regarding their polarities and modes of action. As indi-cated earlier, only in the past several years has the electrophysiology of the worm begun to be examined. A complete connectivity map for the worm's nervous system at the ganglion level, however, had been con-structed earlier and is encoded in a computerized database (Achacoso and Yamamoto 1992; Cherniak 1994). In addition, a study by Raizen and Avery (1994) employed a newly developed method for recording currents, producing what they term "electropharyngeograms," to in-fer specific neuronal effects.

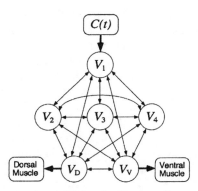

Figure 10.1 Neural network model for the chemotaxis control circuit of *C. elegans*. The state variable of each neuron (circle) is voltage (V_i). The model contains one chemosensory neuron (V_1), three interneurons (V_2–V_4), and two motor neurons (V_D, V_V). The chemosensory neuron receives input equal to the chemical concentration $C(t)$ at the tip of the nose, and the motor neurons innervate dorsal (D) and ventral (V) neck muscles. (From Ferrée and Lockery 1999, 267.)

Two Contrasting Research Programs

Explanations of the behavior of an organism, in this case *C. elegans*, can, as suggested, be pursued at several different levels and from different perspectives. The neural circuits mentioned previously represent one level of explanation, but worm researchers are also interested in more fine-grained mechanisms. A persistent feature of Sydney Brenner's legacy has been the search for individual genes that are strongly related to behaviors (de Chadarevian 1998; Schaffner 1998a; Ankeny 2000). Bargmann's laboratory has investigated several of these genes, particularly those affecting chemosensitivity and chemotaxis. In 1998, she and de Bono published in *Cell* a paper on a genetic explanation for worms' "social" behavior that attracted wide attention, including a *New York Times* article. It is to this essay that I turn next. As an interesting contrast to this genocentric perspective, I compare it with a recent paper from Lockery's laboratory that virtually ignores gene-based explanations in favor of a complex neurophysiological account.

Does the Worm Have a Social Life? A Genetic Basis for Social Behavior. In their 1998 paper de Bono and Bargmann summarized their results as follows:

Natural isolates of *C. elegans* exhibit either solitary or social feeding behavior. Solitary foragers move slowly on a bacterial lawn and disperse across it, while social foragers move rapidly on bacteria and aggregate together [or "clump"]. A loss-of-function mutation in the *npr-1* gene, which encodes a predicted G protein–coupled receptor similar to neuropeptide Y receptors, causes a solitary strain to take on social behavior. Two isoforms of NPR-1 that differ at a single residue occur in the wild. One isoform, NPR-1 215F, is found exclusively in social strains, while the other isoform, NPR-1 215V, is found exclusively in solitary strains. An NPR-1 215V transgene can induce solitary feeding behavior in a wild social strain. Thus, isoforms of a putative neuropeptide receptor generate natural variation in *C. elegans* feeding behavior. (de Bono and Bargmann 1998, 679)

This paper makes strong claims involving a genetic explanation of behavior. At the end of the introduction, the authors write that they "show that variation in responses to food and other animals in wild strains of *C. elegans* is *due to* natural variation in *npr-1*" (1998, 679; emphasis added). The phenotype difference is actually somewhat complex, and it is not just related to social or solitary feeding in the presence of a sufficient bacterial food supply. The social and solitary strains also differ in their speed of locomotion (also termed "hyperactivity"), with solitary wild strains moving more slowly than social strains. Finally the two types differ as well in burrowing behavior in agar. But de Bono and Bargmann contend that "a single gene mutation can give rise to all of the behavioral differences characteristic of wild and solitary strains" (1998, 680).

The authors obtained from three other investigators chemically induced mutants of normally solitary strains that had converted to social feeding behavior. Mapping analysis suggested that all three alleles were due to modifications in a single gene on the X chromosome known as *npr-1*. This gene was then cloned and sequenced and used to rescue the social/hyperactivity phenotype in solitary strains. The difference between social and solitary strains was traced to a single amino acid substitution, phenylalanine for valine, at position 215 in the NPR protein, this due to a single nucleotide G → T mutation. The social behavior mutation is recessive. Sequence comparison with the *C. elegans* database indicated that *npr-1* coded for a seven–transmembrane domain receptor, expressed in neurons, that resembled members of the G protein–coupled neuropeptide Y receptor family. This family can be thought of as a set of similar proteins that can control the activity of other molecules (e.g., ion channels, second messengers, or enzymes) through a chain of causal influences.

De Bono and Bargmann offer several "different models that could explain the diverse behavioral phenotypes of npr-1 mutants." One possibility was that npr-1 might repress a swarming response by inhibiting sensory neurons. Alternatively food could stimulate production of a molecule that binds to the receptor, and as the amount of food decreased the swarming response might be derepressed. The point to be made here, as the authors write, is that "resolution of these models awaits identification of the cells in which npr-1 acts, and the cells that are the source of the npr-1 ligands" (1998, 686).

Thus what we have in this account is a genetic difference (one nucleotide) that "causes" the difference between social and solitary behavioral phenotypes. This notion of cause is one that I believe can best be understood as a factor that is "necessary in the circumstances." The "circumstances" are a huge ceteris paribus clause that assumes organism identity in other relevant aspects. Moreover, the causal pathway(s) relating the gene product (the neuropeptide receptor) and the behavior is not yet known—that is, the function of the models has yet to be resolved. So the link between gene and behavior jumps over not only how the gene produces the tertiary structure of its product (fair enough) but also the nature of the receptor-cell interaction and cell-to-cell causal influences in a neural net leading to locomotion and clumping. Yet the claim has been made—not only in the paper per se but in popular commentary on it by a prominent journalist—that here we find a "spectacular" instance of "genes that govern certain behaviors in animals" (Wade 1998). This G-P arcing, or jumping from gene to phenotype and imputing a causal explanation of the phenotype to a single gene, has been noted before as a problem for reductions or complete molecular explanations (see Kitcher 1994, 395, and also my comments in Schaffner 1998a, 239).

Modeling Worm Circuits and Neurons Physiologically. An interesting contrast with the foregoing account is found in a recent essay by Ferrée and Lockery. Whereas de Bono and Bargmann focus on a DNA nucleotide change as the "cause" of a behavioral phenotype, Ferrée and Lockery provide an analysis that attempts to model the factors and interactions that govern the neurons (Lockery 2000). Their task was to determine "the behavioral strategy for chemotaxis in C. ele-

gans." Their approach is to "derive a linear neural network model of the chemotaxis control circuit" in *C. elegans* and then to "demonstrate that this model is capable of producing nematodelike chemotaxis" (Ferrée and Lockery 1999, 263).

Ferrée and Lockery used an idealization of the candidate neural network proposed by Bargmann (unpublished) (figure 10.1). Lockery's own investigations (Goodman et al. 1998) have shown that the neural signals in *C. elegans* are encoded by graded electrical potentials (not by classic sodium action potentials). The individual neurons display nonlinear transfer functions, but Ferrée and Lockery propose that one can look to a simplified linearization of the chemotaxis system for some insights into this behavior, albeit as a first approximation.

The model building proceeds from the organismal level down. It starts from a model of the nematode body that captures the head and neck turning movements (head-sweep), then seeks a neural implementation of the head-sweep mechanism. Ferrée and Lockery argue that their neural model is based on the worm's neurophysiology, but only— at this point—weakly on its neuroanatomy. Citing Goodman et al. (1998), they suggest that the neurons can be represented as single electrical compartments. (Compartment models are one of the traditional strategies used in neuroscience; see Bower and Beeman 1995; Koch 1998.) An equation for the voltage V_i of the *i*th neuron can be written, using standard compartment modeling, as

$$C_i^{\text{cell}} dV_i/dt = -G_i^{\text{cell}}(V_i - E_i^{\text{cell}}) - I_i^{\text{elec}}(V) - I_i^{\text{chem}}(V) - I_i^{\text{sens}}(t) \qquad (1)$$

where C_i^{cell} is the whole-cell capacitance, G_i^{cell} is the effective ohmic conductance associated with the linear region of the I-V curve, and E_i^{cell} is the resting potential of an isolated neuron. Here $I_i^{\text{elec}}(V)$ and I_i^{chem} represent electrical and chemical synaptic currents, $V = (V_1, \ldots, V_N)$ is an *N*-dimensional vector composed of the voltages of all *N* neurons in the network, and $I_i^{\text{sens}}(t)$ represents chemosensory input (Ferrée and Lockery 1999, 268).

Then, borrowing from data on *Ascaris* (as already noted, synaptic neurophysiological data are not yet available for *C. elegans*), the chemical synapses between cells *i* and *j* can be modeled by the sigmoidal functional equation

$$I_i^{\text{chem}}\,(V) = \sum_{j=1}^{N} G_{ij}^{\text{chem}}\sigma[\beta_{ij}\,(V_j - V_j)](V_i - E_{ij}) \qquad (2)$$

where G_{ij}^{chem} is the maximum conductance in the cell i due to synaptic connections from cell j, and E_{ij} is the reversal potential for the corresponding postsynaptic current. Electrical synapses are similarly modeled by a slightly simpler third equation. Furthermore, chemical inputs to the system are captured by

$$I_i^{\text{sens}}(t) = -\delta_{i1}\kappa_{\text{sens}}C(t) \qquad (3)$$

where $C(t)$ is the chemical concentration at the tip of the worm's nose, δ_{i1} is the standard Kronecker delta, and κ_{sens} is a constant parameter.

The total synaptic model can be further simplified by representing only the chemical synapses. Equation (2), which is then governing, is nonlinear, but it can be linearized by using a Taylor series expansion and retaining only the linear terms. This process yields the following set of equations:

$$dV_i/dt = \sum_{j=1}^{N} A_{ij}Vj + b_i + c_i(t) \qquad (4)$$

(The matrix A_{ij} and constant vector b_i are complicated functions of the Gs, Vs, and Es introduced in equations (1)–(2) and are not reproduced here; see Ferrée and Lockery 1999, 269.) This linearized equation and two quite simple body model equations are then combined with an equation representing the chemical environment C, and the equations are solved to yield a state trajectory $S(t)$ that begins from some specified initial state S_0. The simulation solutions were obtained by numerical integration, and some other tricks were employed to eliminate transients. Figure 10.2 shows a comparison between real and simulated worms.

Next Ferrée and Lockery explored the linearized equation solution to develop a more intuitive result, since they note that "distributed representations" often lack this property of intuitability. This part of their paper provides a "simple rule for chemotaxis control which relates the body rate of turning . . . to time derivatives of the chemo-

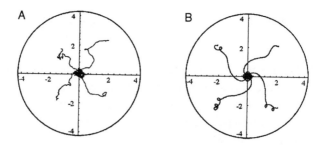

Figure 10.2 Chemotaxis control by neural networks: (A) biological nematodes, (B) model nematodes controlled by network Γ_1. In both cases simulation time $T = 5$ min. Axis labels are in centimeters. (From Ferrée and Lockery 1999, 270.)

sensory input" (1999, 272). Based on the analysis, Ferrée and Lockery argue that their network uses strategies of both klinotaxis (alignment with a vector component of the stimulus field) and klinokinesis (change in turning rate in response to the scalar value of a stimulus field) to produce the behavior represented in figure 10.2B. (Here the definitions of klinotaxis and klinokinesis follow Dunn 1990.) These strategies also suggest seeking additional experimental worm stimulus and movement data to confirm or disprove the models.

Ferrée and Lockery's approach does not rely on genes, and it does not employ structural data from molecular biology. It does utilize physiology and neuroscientific compartment analysis to formulate a mathematical model of a neural network that qualitatively agrees with the worm's observed behavior. It is perhaps more similar to a biophysical approach than a biomolecular one.

Extrapolation from the Worm

What might the worm tell us about other organisms, including ourselves? According to Scott Gilbert and Erik Jorgensen, the answer is "not much." Gilbert and Jorgensen, in commenting on an earlier paper of mine (Schaffner 1998a), indicated that we would not learn anything about human consciousness or agency from so simple an organism as *C. elegans* (Gilbert and Jorgensen 1998, §6). I agree that worm studies will not tell us anything about consciousness or intention or agency, for complexities exist in humans that are not found in simpler organisms. And some of these complexities will include the interaction of linguistic and sociocultural factors with biological developmental processes (see Deacon 1997 for an elaboration of such a view). But some fundamental mechanisms, including simplified analogues of real biological neural nets, are emerging in *C. elegans* studies (see my references to Lockery's research program and also Wicks et al. 1996). Furthermore, it appears that "the molecules and mechanisms of neurogenesis are phylogenetically conserved among worms, flies, and vertebrates" (Fitch and Thomas 1997, 831, with supporting references).

It was only recently that the complete sequence for the *C. elegans* genome was announced in a special issue of *Science*. In that issue, a number of articles reflected on the extrapolation question. In perhaps one of the most interesting comments, that by Chervitz et al. (1998), the authors maintained that whereas yeast (*Saccharomyces cerevisiae*) can be examined for clues to "core functions" (such as DNA and RNA replication and intermediate metabolism) that are common to all cells, *C. elegans* is a source of information for signal transduction that may well apply to all multicellular metazoans. In their abstract they write:

Comparative analysis of predicted protein sequences encoded by the genomes of *Caenorhabditis elegans* and *Saccharomyces cerevisiae* suggests that most of the core biological functions are carried out by orthologous proteins (proteins of different species that can be traced back to a common ancestor) that occur in comparable numbers. The specialized processes of signal transduction and regulatory control that are unique to the multicellular worm appear to use novel proteins, many of which re-use conserved domains. Major expansion of the number of some of these domains seen in the worm may have contributed to the advent of multicellularity. The proteins conserved in yeast and worm are likely to have orthologs throughout eukaryotes; in contrast, the proteins unique to the worm may well define metazoans. (Chervitz et al. 1998, 2022)

These authors also claim that the recent information provided by the worm strongly supports the model organism approach:

The basic assumption that the so-called "model organisms" will provide reliable functional annotation for the human DNA sequence is strongly supported by our observations. First, the sum of the biology of worm and yeast can be obtained efficiently by studying core functions largely in yeast and signal transduction largely in the worm, with virtually no overlap. Second, the evolutionary distance (and biological diversity) between yeast and worm did not interfere with the finding of orthologs and shared domains, making it likely that a robust chain of annotation is possible through all of the eukaryotes. (Chervitz et al. 1998, 2027)

But extrapolation from the worm—even to other organisms simpler than humans—is not without controversy. One debate currently underway raises questions about the phylogenetic position of C. elegans—a position that is being re-evaluated on the basis of molecular sequence information. As Blaxter (1998) points out, traditional morphological criteria placed the worm distal to the branching off of the fly and vertebrates, but the ongoing re-evaluation suggests that Drosophila and C. elegans are more alike to one another, with the arthropods branching off the nematodes (see dendrograms in Blaxter's figure 1, p. 2042). Thus sequences common to two of our major model organisms, the worm and the fly, may not extrapolate well to mice and humans even though they will inform each other. I revisit this issue of phylogeny and extrapolation later in the chapter.

On to the Fly: Drosophila melanogaster Courtship and Mating Behavior as a Model for Behavioral Genetics and Neuroscience

The second "simple" organism I discuss, about which we know a considerable amount regarding genetic mutations that affect behavior, is the common fruit fly, Drosophila melanogaster (or Drosophila for short, when it is not important to distinguish species differences). This organism is the same one that T. H. Morgan and countless other geneticists have investigated since the early twentieth century to explore the relations between genes and phenotypes. As with C. elegans, Drosophila is another prototypical organism whose 120-Mb nucleotide genome is being sequenced as part of the human genome project. For the purposes of this chapter, I concentrate on the complex courtship and mating behavior of Drosophila and on what has recently become known about the cellular and biochemical etiology of aspects of its behavior patterns in this domain.

Though Drosophila has a long history as a model organism, dating to about 1910, our story begins in the 1960s, when Seymour Benzer, a molecular biologist, began to identify behavioral mutants of Drosophila (Greenspan 1995; Weiner 1999). In the 1970s and 1980s, a number of investigators, especially Jeffery C. Hall at Brandeis, continued this work and used a technique involving genetic mosaics that permits the creation of male cells in localized regions of female flies.[2] Hall (1994) summarized his decades-long investigation in a now-classic review article in Science. Greenspan and co-workers' 1995 related research was also reported in *Science* (Ferveur et al. 1995) and in a simplified version in the April 1995 issue of *Scientific American*, which offers an accessible summary of some of the mosaic technique's recent results.

We begin by noting the extraordinary range of behavioral mutants in *Drosophila*, many with colorful and evocative names, such as *couch potato, nerd, fruitless* (or *fru*), and *ether-à-go-go*. (The last name, coined appropriately enough in the 1960s, is based on a shaking-limb mutational behavior produced by the anesthetic.) Table 1 in Hall's (1994) review article summarizes these mutant types and includes as well some additional classical mutations, such as *white eye*.

Hall provides a most interesting account of the species-specific courtship-mating behavior in flies, from about which it is worth quoting at length:

Courtship in *Drosophila melanogaster* involves a series of behaviors, most of which were caught on film during the production of a certain blue movie. [See Hall 1994, 1703, for a series of stills from the "blue movie," which cannot be reproduced here because they must be seen in full color.] Once the male and the female have come into some reasonable proximity (perhaps on a food source or when experimentally put together in a mating cell), they quickly sense each other. Primarily, this seems to be the male detecting the female by using more than one sensory modality. Soon after the male reveals that he has noticed the female (which one infers by observing the orientation of his body toward hers), he taps the female's abdomen. If she is walking about, he follows her during most of the time that she is moving in this manner (no courtship occurs in flight, unlike the capabilities of some other dipterans). As the male orients to a stationary female—including circling her—or follows a mobile one, he frequently sticks out one wing or the other. This extension of the wing is accompanied by its vibration, which produces a "love song" that can be recorded with specialized microphones; the measurable components of these sounds are among the more salient species-specific elements of fruit-fly courtship. Several seconds to a few minutes after the two flies have begun to interact, the male extends his proboscis and licks the female's genitalia. Licking is almost immediately followed by the male's first copulation attempt, which involves an abdominal bending by the male; this can be viewed in more contorted form by looking at the posture accompanying copulation per se. . . .

If an attempted copulation fails, the male may cease courting for some moments. Thus, overt courtship interactions occur only about 60 to 80% of the time when the male and female are together (called the Courtship Index). . . . When the male resumes courting, he almost always drops back to the orientation and following or singing stages (that is, not to tapping or licking) and continues through the rest of the sequence. This series of actions and inter-fly interactions is successful in more than 90% of short-term laboratory observations of wild-type pairs. "Success" means copulation, which has a species-specific duration (about 20 min in *Drosophila melanogaster*). (Hall 1994, 1702)

The "genetic explanation" for this complex choreography is incomplete, but some important mutants that exhibit alternative courtship behavior have been identified and cellular explanations sketched for them. On the basis of these, aspects of the genetic explanation for such behavior have been the focus of recent further research as well as considerable speculation about extrapolation. In his account of the genetics of courtship behavior, Hall focuses on the *fruitless* (or *fru*) mutation, writing that the *fruitless* mutation may define a sex determination factor as well as a courtship gene. This mutation is now distinguished into several variants, *fru¹*, *fru²*, *fru³*, and *fru⁴*. The behavioral phenotype of the *fru¹* mutation in males involves vigorous courtship of females but no attempts at copulation; a *fru¹* male never executes the abdominal bending described by Hall. Hall adds that "the most dramatic reproductive anomaly associated with *fruitless* is that the *fru¹* mutant courts another male just as vigorously as he does a female. Moreover, groups of *fru¹* males will snake around a chamber, forming 'courtship chains' [Hall also calls these "conga lines"] in which most individuals are simultaneously courters and courtees."

The *fru¹* males are also mild song mutants and display other anomalies as well. One important finding was that severe forms of this mutation lacked an abdominal bending muscle, known as the muscle of Lawrence (MOL) after its discoverer. This may account in part for the lack of abdominal bending, but *fru²* mutants lacking an MOL do exhibit both abdominal bending and copulatory behavior with females. The uncertainty about the behavioral causation pathway(s) from *fru* is underscored by several other facts reviewed by Hall, who cautions us against accepting the nihilistic view (my term, not his) that "everything is expressed everywhere and affects everything," though he admits that "some of the effects and inferred noneffects are not very clean." He also adds that "if any mutant type in *Drosophila* is a true behavioral one, then this [*fru*] is it" (1994, 1712).

The problem is that the manner in which *fru* might produce its effects is, despite the enormous amount of work already accomplished, still quite uncertain. The work performed in Greenspan's laboratory on the neural location of sex orientation has made some important additional progress. Greenspan and his associates have developed a different kind of gyandromorph that is primarily male but has small, localized areas of female cells. These had their courtship behavior changed: the males of some of these transformed strains courted male flies as well as female flies. (The technique was an application of Brand and Perrimon's work as adapted by Jean-François Ferveur, working as a postdoctoral student in Greenspan's laboratory.) By some clever breeding techniques, they were able to switch localized neuronal cells to a feminized path of development (see Greenspan 1995 for figures).

It must be added that even with the addition of this neuroanatomical dimension, the explanations offered by Hall and Greenspan are still in the somewhat speculative realm. The pathways from genes through neuronal development to behavior are still quite "gappy." Just as we found in the simpler organism *C. elegans,* much more work will be required before we understand all the details of the complex mechanisms underlying behavior. Greenspan (1995) underscores this and adds that learning phenomena (discussed also by Hall but largely ignored by me in the preceding account) add to the complexity. Greenspan writes that "just as the ability to carry out courtship is directed by genes, so, too, is the ability to learn during the experience. Studies of this phenomenon lend further support to the likelihood that behavior is regulated by a myriad of interacting genes, each of which handles diverse responsibilities in the body" (1995, 75–76).

Another significant advance in this area occurred in 1996 when a team of investigators, including Lisa Ryner at Stanford and Jeff Hall, were able to clone the *fru* gene, which appears to be a regulatory transcription factor. Additional work on *fru*'s pathway indicates that it acts downstream in the sex determination pathway of the fly but sits at the head of a branch of a sex orientation and action pathway functioning in the fly's central nervous system (Ryner et al. 1996). Ryner et al. report that "the sex-specific *fru* products are produced in only about 500 of the 10^5 neurons that comprise the CNS" (1996, 1079). They speculate that *fru* is the final regulatory gene in its branch of the sex determination hierarchy and that it may "directly control the expression of downstream genes responsible for governing specific MOL development, sexual orientation, and the behaviors that comprise male courtship" (1996, 1086). The authors note that it is not clear yet whether *fru* directs neurogenesis or differentiation or both processes, though some related experiments on a different earlier pathway suggest that *fru* is likely to control differentiation (1996, 1086).

At the end of their article, Ryner et al. speculate further on extrapolating fly findings to more complex organisms. Although they note that "mammals and flies use unrelated sex determination mechanisms," they write that "there is a variety of evidence in vertebrates, including humans, suggesting that male sexual behavior, including sexual orientation, has a genetic component" (1996, 1089). In support of this, they cite studies and reviews by Breedlove (1994), Hamer's group (Hu et al. 1995), and LeVay (1996). This claim drew a sharp response in a letter to *Science* by four evolution scholars, who asserted that such extrapolations "detract from the value of the work" of Ryner et al. DeSalle et al. (1997) argue that such extrapolation is dependent on the accuracy of one of two assumptions. One is that courtship behavior can occur in only one way regardless of the organism—unlikely! The other is that there is a common ancestor for flies and mammals that displayed courtship behavior influenced by a homologous component; but they point out that "the last common ancestor of *Drosophila* and *Mammalia* was likely a marine invertebrate with external fertilization without courtship behavior" (1997).

So once again phylogeny raises questions regarding extrapolation—a theme we encountered in discussing the worm, and one that seems likely to persist in cases in which evolutionarily distal organisms are used as "models." This problem is less likely to be serious in our final model organism, the mouse, to which I now turn; nevertheless, even there we shall see that there are problems involving simple extrapolations from simpler systems.

Mus musculus: Is This the Sought-After Super Model?

In the second edition of their textbook *Behavioral Genetics: A Primer,* Plomin et al. write that "the common house mouse (*Mus musculus*) has become the most widely used animal in behavioral genetics research" (1990, 263). In part, this is due to the fact that strongly inbred (and thus approximately 98 percent genetically identical) mice could be obtained, allowing good control over genetic variation. (For a history of the mouse as an experimental system, see Ginsburg 1992.) Greenspan et al. point out that the mouse is also "experimental[ly] attractive [because of] . . . its mammalian heritage and its neuroanatomical, physiological, and genetic closeness to humans" (1995, 557). In point of fact, Silver writes that "an astounding finding has been that nearly all human genes have counterparts in the mouse genome, which can almost always be recognized by cross-species hybridization" (1995, 11). In addition, mice display more complex and plastic behaviors than do simpler organisms such as the worm and fly models. In the preface to their collection of papers on mouse genetics, brain, and behavior, Goldowitz et al. offer valuable methodological insights into the choice of an appropriate organism:

To achieve a probing analysis of the genetic foundations of behavior it is important that the species meet three stringent criteria. (1) The organism has a well-documented genetics. (2) The organism demonstrates a behavioral complexity that extends its usefulness to other behaving organisms as a model system. (3) The organism lends itself to a wide range of experimental paradigms. These criteria leave behind rats, bacteria, and humans, considering criteria 1, 2, and 3, respectively. The two organisms that best fit the criteria are the fruit fly and the mouse. (1992, v)

We have already seen the value of the fly questioned on evolutionary grounds. Might the mouse be the Holy Grail of genetic and neuroscientific modeldom?

The popularity of the mouse as a model organism has, if anything, increased in the past few years thanks to powerful genetic manipulation techniques that make this organism especially malleable experimentally. Greenspan et al. (1995) point out that obtaining mutant strains of mice had been difficult before the advent of molecular modification tools. One form of genetic manipulation can produce "knockout" mice that are particularly useful in behavioral geneticists' search for single-gene effects on behavior. In knockouts, single genes can be eliminated and the effects on the mice analyzed. Kandel's and Tonegawa's laboratories have been active in pursuing such models (about which more can be read in the chapter by Craver and Darden in this volume). Tonegawa's laboratory (Chen et al. 1994) has also investigated a calcium-calmodulin kinase II (CaMKII)–deficient mouse strain that displays increased aggressiveness and decreased fear response. (This report generated a semisatirical letter to *Science* by Vogel [1995] on "Wagnerian Genetics," suggesting that a human homologue might account for Siegfried's behavior; it drew a response from Chen.) Another report by Cases et al. (1995), using transgenically modified aggressive mice lacking monoamine oxidase A (an enzyme that degrades serotonin and norepinephrine, two important neurotransmitters), draws genetic parallels with a Dutch human family's behavior as investigated by Brunner et al. (1993; Brunner 1996). Furthermore, using a colony of mice mutants deficient in nitric oxide function, Nelson et al. (1995) reported "a large increase in aggressive behavior and excess, inappropriate sexual behavior in nNOS-mice."

These genetic manipulation techniques in the mouse are discussed in depth in a book by Silver (1995). He writes that the mouse model now allows us to move beyond the search for single-gene effects and pursue an explanation of complex traits "at the molecular level," including those "polygenic differences that control characteristics as diverse as size, life span, reproductive performance, aggression, and levels of susceptibility or resistance to particular diseases, both infectious and inherited" (158). There have been some stunning advances using both knockout and transgenic mice (in which human genes— often putative disease-causing genes—are inserted into the mouse at the embryonic stage).[3] Some of the most interesting mice from a neuroscientific point of view are transgenic mice that develop close analogues of Alzheimer's disease (AD). It is to these exquisitely honed, if forgetful, super models that I turn for my last example.

AD is a devastating neurological disease that primarily affects those of advanced age. It has major effects on memory and on personality, and it is eventually fatal. There are (at least) two subtypes of the disease: the early-onset form, affecting individuals in their late forties and fifties, and the more typical late-onset form, sometimes called familial (sporadic) Alzheimer's disease (FAD), which begins to strike in significant percentages in the seventies. The early-onset form is due to at least three kinds of gene mutations: in APPs, the gene for amyloid precursor proteins (APPs) found on chromosome 21; in the Presenilin 1 and Presenilin 2 genes, found on chromosomes 14 and 1, respectively; and in the apolipoprotein E (ApoE) gene, found on chromosome 19. ApoE has three allelic forms (E2, E3, and E4), and it is ApoE4 in a double dose that is the bad actor. The early-onset forms of AD involve dominant genes of high penetrance, whereas ApoE4's contribution to the more typical late-onset form of AD is probabilistic and

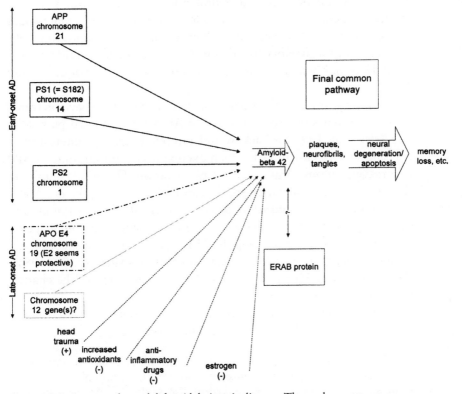

Figure 10.3 A network model for Alzheimer's disease. The early-onset genes (APP, PS1, and PS2) are deterministic (each is sufficient in humans), whereas APO E4 is probabilistic or contributory. The role(s) of the gene(s) on chromosome 12 is still speculative (Scott et al. 1999). Possible environmental influences representing increased risk (+) and decreased risk (–) are shown toward the bottom of the figure. Interactions of the ERAB protein with amyloid-beta 42 are speculative (Yan et al. 1997). These various influences are believed to result in a final common pathway as shown (Lendon et al. 1997; Yan et al. 1997).

best characterized as a "risk factor" increasing the likelihood of the disease in some patients (those who are E4/E4) by five- to fifteenfold. See figure 10.3 for a graphic account of these genes' influences.

AD's pathology is increasingly well understood. Autopsies of these patients show neurofibrillary tangles (NFTs) and senile plaques in their brain tissue. The tangles display tau epitope immunoreactivity. The plaques contain deteriorated neural cells associated with what are termed Aβ peptides. These Aβ peptides of about 4K molecular weight are remnants of APP proteins, which seem to be required in their normal form for memory formation. A subform of Aβ peptide known as Aβ42 is highly neurotoxic and is deposited early in plaques (figure 10.3).[4]

In recent years several investigators have been able to develop transgenic mice that show several forms of Aβ peptide–related pathology (see Price et al. 1998 for background and references). Of interest to us is Hsiao's transgenic model, which overexpresses mutant human APP in transgenic mouse brains in the Tg2576 strain. The brains of mice in this line, as Price et al. write, show "elevated levels of Aβ40 and Aβ42; dystrophic neurites; and Aβ deposits in amygdala, hippocampus, and cortex. The Tg2576 mice have shown impairments at a young age on several memory tests including the Morris water maze, a spatial reference memory task, and the Y-maze alternation tasks" (1998, 1081). (For more about Morris water mazes and other tests, see the chapter by Craver and Darden in this volume.)

Price et al. conclude their account of this model and other mouse models with the following exciting summary:

Multiple studies in transgenic mice provide strong evidence to support the view that Aβ42 formation is an early and critical pathogenic event: mice expressing APP with mutations linked to disease develop Aβ deposits at an accelerated pace, coexpression of PS variants linked to FAD accelerates the rate of Aβ deposition, and ApoE plays a role in modulating the rate of Aβ deposition. Thus, three known genetic causes or risk factors for FAD and AD effect Aβ deposition. Whether therapeutic agents that affect the concentration, deposition, aggregation, degradation, clearance, or toxicity of Aβ will influence the clinical and pathological features of AD is unclear. Nevertheless, it seems likely that approaches that reduce the concentration of Aβ or the rate of amyloid aggregation and deposition in proximity to synapses and neuronal cell bodies could be beneficial for patients with AD. Clearly, as brain penetrant agents with these activities are developed, the aforementioned transgenic models will be indispensable for validating the in vivo efficacy of these agents. (Price et al. 1998, 1081)

So mice are evolutionarily close to us, and these recent gene knockout and gene insertion techniques make them superb models for investigating neuroscientific processes and pathology. Are there any limitations to what these super models can help us accomplish?

There are a few caveats in the literature. A model is a model, not an identity relationship. And even in the aforementioned APP model, mice do not develop AD *exactly* as humans do. As Price et al. (1988, 1081) note, "transgenic models of Aβ amyloidogenesis do not fully recapitulate the entire spectrum of neuropathological alterations seen in human AD cases. For example, although phosphorylated tau epitopes are present in dystrophic neurites, NFTs have not been described.[5] Moreover, neuronal loss is only observed in mice with extremely high amyloid burdens." So the models are extraordinary—but not perfect.

There are also some general concerns about the transgenics and knockouts that suggest caution in the interpretation of data obtained from these genetic manipulations. Generally these caveats cite the roles of incomplete controls, genetic backgrounds, and pleiotropic effects of manipulations on the phenotypes of interest. A remarkable exchange initiated by Robert Gerlai (Gerlai et al. 1996) indicates the difficulties that can affect research in this area. Gerlai concisely pointed out that (1) polymorphisms in the genetic background can have significant effects, (2) null-mutant mice are often *hybrids* of two mouse strains, (3) these hybrids are genetically different from their control littermates at loci other than the gene of interest, and (4) the widespread use of the 129 type strain would have the result that alleles surrounding the targeted gene would be of the 129 type in the null-mutants, whereas they would be of the BL6 type in the wild-type controls. Further, it would take many generations of breeding to correct for these potential confounders. Gerlai's cautions were for the most part supported by accompanying comment articles by Crawley, Crusio, and Lathe.

In a subsequent review article discussing the wide-ranging use of inbred mouse strains and their implications for molecular studies, Crawley et al. reiterate these concerns, writing that "background genes from the parental strains may interact with the mutated gene, in a manner which could severely compromise the interpretation of the mutant phenotype" (1997, 107). Thus, unless meticulous care is taken in choosing the proper inbred strain to work with, there are background effects that may mask the sought-after effect. Gerlai and Clayton (1999) cite other factors that must be considered and controlled to avoid confounding factors. He argues that the ethological perspective is often lacking in laboratory scientists, with serious results. For example, mice are not well adapted to water, whereas rats and some other species of rodents are; so performance in a Morris water maze will drastically differ across species. Moreover, "exploratory behavior," which is what is typically tested in animal models in search of genetic influences, can be influenced by hunger, sex drive, fear, and chronic stress, all of which must be controlled. Attribution of a spatial learning defect to a molecular feature (e.g., Silva and co-workers' claim that the cause is the lack of the alpha unit of CaMKII) may be incorrect; the defect may actually be due to a fear response caused by laboratory handling (Gerlai and Clayton 1999, 50).

Crusio (1999), reporting on a November 1998 meeting on knockouts and mutants, notes that gene knockout interpretation is under close scrutiny by the field's leaders, and that some are urging increased test standardization to avoid errors. This is a view to which Crusio takes some exception. But he does argue that, though current strategies may identify general mechanisms related to "monomorphic genes identical in all individuals" (100), they will not address individual variations in task performance well, unless they are supplemented. Similar concerns have been echoed in the history and philosophy of science literature by Perlman (1997) and Culp (1997).

Most recently, a remarkable paper by Crabbe et al. (1999) reported an attempt to compare behaviors of several inbred mouse strains in three different laboratories. Major efforts were made to standardize the three laboratories' environments, but in spite of this significant differences in behaviors were found. "Strains differed markedly in all behaviors, and despite standardization, there were systematic differences in behavior across labs. For some tests, the magnitude of genetic differences depended upon the specific testing lab. Thus, experiments characterizing mutants may yield results that are idiosyncratic to a particular laboratory" (Crabbe et al. 1999, 1670). This article drew several comments from other investigators, which were published in a subsequent issue of *Science* (Picciotto and Self, and others, 1999), but mysteries about the sources of the variations in behavior remain, and this debate will continue.

From a philosophical point of view, perhaps the most interesting comment in this debate about the mouse model was a diagnosis of *the real problem* by Gerlai. In his reply to the comments on his 1996 paper he wrote that most of "the problems arise from a deeper theoretical root"—that it is "really a systemic one that concerns biological organization and the functional units of this organization." He added:

From a geneticist's viewpoint, the units of biological organization are clearly the genes and their function is to encode particular proteins. However, one might argue that when it comes to the question of phenotypical effects, genes might not be the "units" and the definition of their function might be more complicated. One might suggest that groups of genes defined by higher–organizational level phenomena, including developmental, physiological, or even behavioral phenomena, might represent the functionally relevant unit. Disruption of a single gene might alter biochemical processes within the functional gene group. Expression levels of the gene might change in concert. (Gerlai et al. 1996, 189)

Gerlai adds that the boundaries of these gene groups might not be sharp, and that some genes might belong to more than one functional group.

These are themes that have been under vigorous discussion in the more philosophical literature, usually under the heading of "developmental systems theory" and the "developmentalist challenge" to more traditional approaches to genetics. Elaborating on them within the context of the model organisms discussed here would require at least another chapter. To some extent, I have addressed the developmentalist perspective in connection with C. *elegans* in other articles (Schaffner 1998a, 1998b), and Griffiths and Knight's (1998) comments provide a more developmentalist view as well.

Summary and Conclusion

This chapter has reviewed several model organisms from the perspective of the question of extrapolation from organism to organism and also to humans. I have had to be selective with respect to both organism and area of investigation within those organisms. One defect of this chapter is that I have not addressed any primate models—but there the genetics, which is likely to give us the best initial purchase on mechanisms, is not well developed. I have also focused my attention on courtship behavior in *Drosophila* rather than on memory formation, as addressed in Tim Tully's (1996) important work partly for historical reasons and partly because of the broad interest of the *Drosophila* community in this area. The mouse is more genetically tractable than the rat, whence my choice of *Mus musculus,* but even there the field includes many different kinds of investigations. I have focused on AD because its genetics and pathophysiology are fairly well known.

Philosophically, I think the take-home lesson from these examples is that animal models are ultimately used as a source for extrapolations to humans, though an important subordinate goal is to understand the biology of all organisms. One can experiment more easily in simple systems, and in addition the ethical constraints on human experimentation are more restrictive than those on animal research (though some might disagree), making some interventions impossible in humans. The great hope is that simple systems will disclose features that are *broadly conserved* throughout other species and that can be taken as the basis for a science of behavior, learning, and consciousness. Another hope—or better, assumption—is that simpler organisms will disclose more easily understood *simple* mechanisms.[6] But as we have seen in my concluding comments about the mouse model, an awareness of the importance of complexity beyond single gene (or allele)–single trait models is becoming more pronounced. Contemporary biology has its fingers crossed that simple mechanisms are the ones that are broadly conserved, and that biology research programs can simultaneously aim at a kind of simplicity and universalizability. Good luck to contemporary biology!

Biology and the biomedical sciences will need good luck, as well as laborious efforts, because the phylogenetic debates about some of the organisms I have discussed are already raising concerns about extrapolatability to mammals from worms and flies. As stressed toward the end of this chapter, the methods used in the design of experiments with animals that ignore genetic background and environmental influences may result in misattribution of an effect to a gene or circuit mechanism or in missing the true mechanisms altogether. Recognition of the limitations is, however, the first step toward a more robust set of inquiries that can address these problems and, one hopes, discover mechanisms in animal models that extrapolate to life in general and to humans in particular.

NOTES

The research leading to this chapter was partially supported by the National Science Foundation's Studies in Science, Technology, and Society Program. I thank Shawn Lockery and his colleagues for sharing their research in progress with me. Parts of the first and second sections rely on Schaffner (2000) and parts of the third section rely on Schaffner (2001).

1. Bargmann (1993, 49) quotes figures from Durbin (1987): "For any synapse between two neurons in any one animal, there was a 75% chance that a similar synapse would be found in the second animal . . . [and] if two neurons were connected by more than two synapses, the chances they would be interconnected in the other animal increased greatly (92% identity)."

2. This technique is based on the fact that in *Drosophila* the sex of a cell is based on its sex chromosomes: a double X is a female cell and a single X is a male cell. It is possible to make X chromosomes unstable, such that during mitosis a clone of the female parent cell becomes a male cell. This can occur at several points in *Drosophila* development, and when it takes place at late stages it can leave small islands of male cells in an otherwise female sea. These mixed-sex (at the cellular level) organisms are called gynandromorphs (Lawrence 1992, 10, 82–87).

3. Transgenic manipulation can also be accomplished in the fly. For an overview of both fly and mouse transgenic and knockout methods, see Alberts et al. (1994, 327–30).

4. For an excellent introduction to the pathology see Price et al. (1998, esp. 1080–81); for accounts of various findings in related animal models see the "Online Mendelian Inheritance in Man" article on APP at http://www.ncbi.nlm.nih.gov/htbin-post/Omim/dispmim/104760 #ANIMAL MODEL.

5. Lee and Trojanowski (1999) also raise some questions about APP (though they do not rule out its significance) and suggest that another gene (tau) may be more important in the neurodegeneration seen in AD models in mice.

6. There is a growing awareness that even our *simple* mechanisms are really not so simple, and that we bias our understanding of how they work by selective attention to only some of the interacting components of mechanisms. An excellent example of just how complex the postsynaptic neuronal response is in actuality is provided by Weng et al. (1999). This issue of complexity, however, is not one that can be addressed within the confines of this chapter.

REFERENCES

Achacoso, T. B., and W. S. Yamamoto. 1992. *AY's Neuroanatomy of* C. elegans *for Computation*. Boca Raton, Fla.: CRC Press.
Alberts, B. 1997. "Preface." In D. Riddle, T. Blumenthal, B. J. Meyer, and J. Priess, eds., C. elegans *II*. Cold Spring Harbor, N.Y.: Cold Spring Harbor Press, xii–xiv.
Alberts, B., D. Bray, J. Lewis, M. Raff, K. Roberts, and J. D. Watson. 1994. *The Molecular Biology of the Cell*, 3rd ed. New York: Garland.
Ankeny, R. 2000. "Fashioning Descriptive Models in Biology: Of Worms and Wiring Diagrams." *Philosophy of Science 67*.
Avery, L., D. Raizen, and S. R. Lockery. 1995. "Electrophysiological Methods." In H. Epstein and D. Shakes, eds., *Modern Biological Analysis of an Organism*. New York: Academic Press, 251–69.

Bargmann, C. 1993. "Genetic and Cellular Analysis of Behavior in *C. elegans.*" *Annual Review of Neuroscience* 16: 47–51.

Barinaga, M. 1994. "From Fruit Flies, Rats, Mice: Evidence of Genetic Influence." *Science* 264: 1690–93.

Blaxter, M. 1998. "*C. elegans* Is a Nematode." *Science* 282: 2041–46.

Bower, J. M., and D. Beeman. 1995. *The Book of GENESIS: Exploring Realistic Neural Models with the GEneral NEural SImulation System.* New York: Springer-Verlag.

Breedlove, S. M. 1994. "Sexual Differentiation of the Human Nervous System." *Annual Review of Psychology* 45: 389–418.

Brunner, H. G. 1996. "MAOA Deficiency and Abnormal Behavior: Perspectives on an Association." In G. R. Bock and J. A. Goode, eds., *Genetics of Criminal and Antisocial Behavior.* Ciba Foundation Symposium 194. New York: Wiley, 155–67.

Brunner, H., M. Nelen, X. Breakefield, H. Ropers, and B. van Oost. 1993. "Abnormal Behavior Associated with a Point Mutation in the Structural Gene for Monoamine Oxidase A." *Science* 262: 578–80.

Cases, O., I. Seif, J. Grimsby, P. Gaspar, K. Chen, S. Pournin, U. Muller, M. Aguet, C. Babinet, J. Shih, and E. DeMaeyer. 1995. "Aggressive Behavior and Altered Amounts of Brain Serotonin and Norepinephrine in Mice Lacking MAOA." *Science* 268: 1763–66.

C. elegans Sequencing Consortium. 1998. "Genome Sequence of the Nematode *C. elegans:* A Platform for Investigating Biology." *Science* 282: 2012–18.

Chalfie, M., and J. White. 1988. "The Nervous System." In Wood 1988, 337–91.

Chen, C., D. Rainnie, R. Greene, and S. Tonegawa. 1994. "Abnormal Fear Response and Aggressive Behavior in Mutant Mice Deficient for a-calcium-calmodulin kinase-II." *Science* 266: 291–94.

Cherniak, C. 1994. "Component Placement Optimization in the Brain." *Journal of Neuroscience* 14: 2418–27.

Chervitz, S. A., L. Aravind, G. Sherlock, C. A. Ball, E. V. Koonin, S. S. Dwight, M. A. Harris, K. Dolinski, S. Mohr, T. Smith, S. Weng, J. M. Cherry, and D. Botstein. 1998. "Comparison of the Complete Protein Sets of Worm and Yeast: Orthology and Divergence." *Science* 282: 2022–28.

Cook-Deegan, R. 1994. *Gene Wars.* New York: Norton

Crabbe, J. C., D. Wahlsten, and B. C. Dudek. 1999. "Genetics of Mouse Behavior: Interactions with Laboratory Environment." *Science* 284: 1670–72.

Crawley J. N., J. K. Belknap, A. Collins, J. C. Crabbe, W. Frankel, N. Henderson, R. J. Hitzemann, S. C. Maxson, L. L. Miner, A. J. Silva, J. M. Wehner, A. Wynshaw-Boris, and R. Paylor. 1997. "Behavioral Phenotypes of Inbred Mouse Strains: Implications and Recommendations for Molecular Studies." *Psychopharmacology* 132: 107–24.

Crusio, W. 1999. "Using Spontaneous and Induced Mutations to Dissect Brain and Behavior Genetically." *Trends in Neuroscience* 22: 100–102.

Culp, S. 1997. "Establishing Genotype/Phenotype Relationships: Gene Targeting as an Experimental Approach." *Philosophy of Science* 64: S268–78.

Deacon, T. W. 1997. *The Symbolic Species: The Co-evolution of Language and the Brain.* New York: Norton.

De Bono, M., and C. Bargmann. 1998. "Natural Variation in a Neuropeptide Y Receptor Homolog Modifies Social Behavior and Food Response in *C. elegans.*" *Cell* 94: 677–89.

De Chadarevian, S. 1998. "Of Worms and Programmes: *Caenorhabditis elegans* and the Study of Development." *Studies in History and Philosophy of Biology and Biomedical Sciences* 29C: 81–105.

DeSalle, R., R. Bang, M. Yudell, and R. Meier. 1997. "Predictive Value of *Drosophila.*" *Science* 275: 1401–4.

Dunn, G. A. 1990. "Conceptual Problems with Kinesis and Taxis." In J. P. Armitage and J. M. Lackie, eds., *Biology of the Chemotactic Response.* Cambridge: Cambridge University Press, 1–14.

Durbin, R. M. 1987. "Studies on the Development and Organization of the Nervous System of *Caenorhabditis elegans.*" Ph.D. dissertation, Cambridge University, U.K.

Ferrée, T. C., and S. R. Lockery. 1999. "Computational Rules for Chemotaxis in the Nematode *C. elegans.*" *Journal of Computational Neuroscience* 6: 263–77.

Ferveur, J.-F., K. Störtkuhl, R. Stocker, and R. Greenspan. 1995. "Genetic Feminization in Brain Structures and Changed Sexual Orientation in Male *Drosophila.*" *Science* 267: 902–5.

Fitch, D. H. A., and W. K. Thomas. 1997. "Evolution." in D. Riddle, T. Blumenthal, B. J. Meyer, and J. Priess, eds., *C. elegans II.* Cold Spring Harbor, N.Y.: Cold Spring Harbor Press, 815–50.

Gerlai, R., and N. Clayton. 1999. "Analysing Hippocampal Function in Transgenic Mice: An Ethological Perspective." *Trends in Neuroscience* 22: 47–51.

Gerlai, R. 1996. "Gene-Targeting Studies of Mammalian Behavior: Is It the Mutation or the Background Genotype?" *Trends in Neuroscience* 19: 177–81.

Gilbert, S. F., and E. M. Jorgensen. 1998. "Wormholes: A Commentary on K. F. Schaffner's 'Genes, Behavior, and Developmental Emergentism.'" *Philosophy of Science* 65: 259–66.

Ginsburg, B. 1992. "Muroid Roots of Behavior Genetic Research: A Retrospective." In Goldowitz et al. 1992, 3–14.

Goldowitz, D., D. Wahlsten, and R. Wimer, eds. 1992. *Techniques for the Genetic Analysis of Brain and Behavior: Focus on the Mouse.* Amsterdam: Elsevier.

Goodman, M. B., D. H. Hall, L. Avery, and S. R. Lockery. 1998. "Active Currents Regulate Sensitivity and Dynamic Range in *C. elegans* Neurons." *Neuron* 20: 763–72.

Greenspan, R. J. 1995. "Understanding the Genetic Construction of Behavior." *Scientific American* 272(4): 72–78.

Greenspan, R. J., E. Kandel, and T. Jessel. 1995. "Genes and Behavior." In Kandel et al. 1995, 555–77.

Griffiths, P. E., and R. D. Knight. 1998. "What Is the Developmentalist Challenge?" *Philosophy of Science* 65: 253–58.

Hall, J. C. 1994. "The Mating of a Fly." *Science* 264: 1702–14.

Hodgkin, J., R. H. A. Plasterk, and R. H. Waterston. 1995. "The Nematode *Caenorhabditis elegans* and Its Genome." *Science* 270: 410–14.

Hu, S., A. Pattatucci, C. Patterson, L. Li, D. Fulker, S. Cherny, L. Kruglyak, and D. Hamer. 1995. "Linkage between Sexual Orientation and Chromosome Xq28 in Males but Not in Females." *Nature Genetics* 11: 248–56.

Kandel, E. R., J. H. Schwartz, and T. M. Jessell, eds. 1995. *Essentials of Neural Science and Behavior.* Norwalk, Conn.: Appleton and Lange.

Kitcher, P. 1994. "1953 and All That: A Tale of Two Sciences." In E. Sober, ed., *Conceptual Issues in Evolutionary Biology,* 2nd ed. Cambridge, Mass.: MIT Press, 379–99. (Originally published in *Philosophical Review* 93: 335–73 [1984].)

Koch, C. 1998. *Biophysics of Computation: Information Processing in Single Neurons (Computational Neuroscience).* New York: Oxford University Press.

Lawrence, P. 1992. *The Making of a Fly: The Genetics of Animal Design.* Oxford: Blackwell.

Lee, V., and J. Trojanowski. 1999. "Neurodegenerative Tauopathies: Human Disease and Transgenic Mouse Models." *Neuron* 24: 507–10.

Lendon, C. L., F. Ashall, and A. M. Goate. 1997. "Exploring the Etiology of Alzheimer Disease Using Molecular Genetics." *Journal of the American Medical Association* 277: 825–31.

LeVay, S. 1996. *Queer Science: The Use and Abuse of Research into Homosexuality.* Cambridge, Mass.: MIT Press.

Lockery, S. R. 2000. Statement of Research Program. http://chinook.uoregon.edu (which also contains accounts of additional work).

Mann, C. 1994. "Behavioral Genetics in Transition." *Science* 264: 1686–89.

Mori, I., and Y. Ohshima. 1995. "Neural Regulation of Thermotaxis in *Caenorhabditis elegans.*" *Nature* 376: 344–48.

Nelson, R. J., G. E. Demas, P. L. Huang, M. C. Fishman, V. L. Dawson, T. M. Dawson, and S. H. Snyder. 1995. "Behavioural Abnormalities in Male Mice Lacking Neuronal Nitric Oxide Synthase. *Nature* 378: 383–86.

Perlman, R. 1997. "What Transgenic Mice Tell Us about Development." Paper presented at the meeting of the International Society for History, Philosophy, and Social Studies of Biology, July 1997, Seattle (privately circulated).

Picciotto, M. R., and D. W. Self; L. A. Pohorecky; G. R. Dawson, J. Flint, and L. S. Wilkinson; R. Hen; M. G. Tordoff, A. A. Bachmanov, M. I. Friedman, and G. K. Beauchamp; D. Wahlsten, J. Crabbe, and B. Dudek. 1999. "Testing the Genetics of Behavior in Mice." *Science* 285: 2067–70.

Plomin, R., J. DeVries, and G. McClearn. 1990. *Behavioral Genetics: A Primer,* 2nd ed. New York: Freeman.

Price, D. L., S. S. Sisodia, and D. R. Borchelt. 1998. "Genetic Neurodegenerative Diseases: The Human Illness and Transgenic Models." *Science* 282: 1079–83.

Raizen, D., and L. Avery. 1994. "Electrical Activity and Behavior in the Pharynx of *Caenorhabditis elegans.*" *Neuron* 12: 483–95.

Ryner, L. C., S. F. Goodwin, D. H. Castrillon, A. Anand, A. Villella, B. S. Baker, J. C. Hall, B. J. Taylor, and S. A. Wasserman. 1996. "Control of Male Sexual Behavior and Sexual Orientation in *Drosophila* by the *fruitless* Gene." *Cell* 87: 1079–89.

Schaffner, K. F. 1998a. "Genes, Behavior, and Developmental Emergentism: One Process, Indivisible?" *Philosophy of Science* 65: 209–52.

———. 1998b. "Model Organisms and Behavioral Genetics: A Rejoinder." *Philosophy of Science* 65: 276–88.

———. 2000. "Behavior at the Organismal and Molecular Levels: The Case of *C. elegans*." *Philosophy of Science* 67.

———. 2001. "Genetic Explanations of Behavior: Of Worms, Flies, and Men." In D. Wasserman and R. Wachbroit, eds., *Genetics and Criminal Behavior: Methods, Meanings, and Morals*. New York: Cambridge University Press.

Scott, W. K., J. M. Grubber, S. M. Abou-Donia, T. D. Church, A. M. Saunders, A. D. Roses, M. A. Pericak-Vance, P. M. Conneally, G. W. Small, and J. L. Haines. 1999. "Further Evidence Linking Late-Onset Alzheimer Disease with Chromosome 12." *Journal of the American Medical Association* 281: 513–14.

Silver, L. 1995. *Mouse Genetics*. New York: Oxford University Press.

Sulston, J. E., E. Schierenberg, J. G. White, and J. N. Thomson. 1983. "The Embryonic Cell Lineage of the Nematode *Caenorhabditis elegans*." *Developmental Biology* 100: 64–119.

Takahashi, J., L. Pinto, and M. Vitaterna. 1994. "Forward and Reverse Genetic Approaches to Behavior in the Mouse." *Science* 264: 1724–33.

Thomas, J. H. 1994. "The Mind of a Worm." *Science* 264: 1698–99.

Tully, T. 1996. "Discovery of Genes Involved with Learning and Memory: An Experimental Synthesis of Hirschian and Benzerian Perspectives." *Proceedings of the National Academy of Sciences of the USA* 93(24): 13,460–67.

Vogel, H. 1995. "Wagnerian Genetics." *Science* 267: 437.

Wade, N. 1998. "Can Social Behavior of Man Be Glimpsed in a Lowly Worm?" *New York Times*, September 7.

Weiner, J. 1999. *Time, Love, Memory: A Great Biologist and His Quest for the Origins of Behavior*. New York: Knopf.

Weng, G., U. S. Bhalla, and R. Iyengar. 1999. "Complexity in Biological Signaling Systems." *Science* 284: 92–96.

White, J. G., E. Southgate, J. N. Thomson, and S. Brenner. 1986. "The Structure of the Nervous System of the Nematode *Caenorhabditis elegans*." *Philosophical Transactions of the Royal Society, Series B* 314: 1–340.

Wicks, S. R., and C. H. Rankin. 1995. "Integration of Mechanosensory Stimuli in *Caenorhabditis elegans*." *Journal of Neuroscience* 15: 2434–44.

Wicks, S. R., C. J. Roehrig, and C. H. Rankin. 1996. "A Dynamic Network Simulation of the Nematode Tap Withdrawal Circuit: Predictions Concerning Synaptic Function Using Behavioral Criteria." *Journal of Neuroscience* 16: 4017–31.

Wood, W., ed. 1988. *The Nematode* Caenorhabditis elegans. Cold Spring Harbor: Cold Spring Harbor Press.

Yan S. D., J. Fu, C. Soto, X. Chen, H. Zhu, F. Al-Mohanna, K. Collison, A. Zhu, E. Stern, T. Saido, M. Tohyama, S. Ogawa, A. Roher, and D. Stern. 1997. "An Intracellular Protein That Binds Amyloid-Beta Peptide and Mediates Neurotoxicity in Alzheimer's Disease." *Nature* 389(6652): 689–95.

11

Under the Lamppost

Commentary on Schaffner

Marcel Weber
Center for Philosophy and Ethics of Science, University of Hanover, Hanover, Germany

Most of the great biological discoveries of the twentieth century are closely linked to some laboratory organism that, for a variety of reasons, provided scientists with the experimental possibilities necessary to map some genetic, cellular, or molecular mechanism. To mention just a few of the most important examples: The bacterium *Escherichia coli* and its bacteriophages have yielded insights into a large number of complex genetic mechanisms responsible for DNA replication, recombination, mutagenesis, repair, transcription, and translation, and for the regulation of gene expression. The fruit fly *Drosophila melanogaster*—already the favorite animal of classical and population geneticists during the first half of the century—has recently enabled biologists to unravel some of the genetic mechanisms controlling embryonic development. Baker's yeast (*Saccharomyces cerevisiae*) continues to be one of the main organisms of choice for studying the biogenesis of membranes and organelles, intracellular protein sorting, secretion, cell cycle regulation, and other mechanisms, some of which are absent in the prokaryote *E. coli*. The mouse *Mus musculus* was instrumental in advances in immunology that have led to a greatly increased understanding of the body's defense mechanisms against infectious agents and tumor growth.

If neuroscientists increasingly look to model organisms in their attempts to understand the molecular basis of behavior, information processing in the brain, and neuropathogenesis, as Schaffner shows in his chapter, they are simply following a research strategy that has proved to be extremely successful in biology.

All of the discoveries just mentioned share a common feature that distinguishes them from, say, a botanist's elucidation of the life cycle of some rare plant: they are thought to be of *universal* or at least broad significance for the understanding of life in general. A subset of these findings is considered to be especially important because they apply to the human body and could potentially be used to cure diseases. For these reasons, any account of the epistemological foundations of modern biology would be incomplete without an answer to the question of what grounds inferences drawn from some model organism to life in general or to other organisms like human beings.[1]

Historians and philosophers of biology have, to be sure, noticed the widespread use of model organisms in biological research, and there exists a growing body of scholarly work on this subject. However, most of these scholars have looked at model organisms from a different angle. They tend to view model organisms as a part of the "material culture" of experimental biology. Such cultures or technologies can form units of historical analysis in their own right, which can be investigated, to some extent, independently of the development of scientific theories.[2] This also echoes the creed of the "New Experimentalists" that much of the dynamics of scientific change can be understood by looking at experimental systems alone (Hacking 1983, 1992; Rheinberger 1997). Experimental biology rarely conforms to the philosophical stereotype of hypothesis testing, according to which theory always comes first. Biologists perform many experiments simply because they are possible with the experimental system they are using, without having some bold conjecture or specific question in mind. Such experiments are important for increasing scientists' skills in manipulating the experimental systems in a controlled fashion. Specific questions or hypotheses may crop up only once scientists know what their experimental systems can and cannot do.

The fact that biological research is driven, to a large extent, by its experimental systems may be viewed as an important challenge to claims to *universality*, which pervade much of modern experimental biology.[3] Since experimental systems in biology crucially depend on some model organism, an element of *historical contingency* is added to scientific development, which may affect even what theories end up being accepted.[4] Experimental organisms are selected for a variety of reasons, which include short generation time, availability of mutants, and simplicity of genetic system, but also considerations like the ease and cost of maintaining and handling laboratory populations. It may have happened just as often in the history of biology that the choice of organism defined the scientific problems to be solved rather than the other way around (Lederman and Burian 1993, 237). There can be little doubt that the choice of laboratory organisms has influenced the direction of biological research, even if one does not want to go so far as those sociologists of science who have claimed that research tools and the "jobs" they are selected for are "co-constructed" (Clarke and Fujimura 1992, 7).

Another important factor is that the organisms used in experimental studies are to some extent *designed* for this very purpose. Laboratory strains are carefully selected, bred, or even genetically engineered to yield reproducible results. Many experiments are possible only in stabilized, genetically well-defined organisms that possess the requisite markers and that lack certain natural functions that may confound experimental results.[5] Thus the typical laboratory organism differs considerably from its conspecifics living in natural populations.[6]

In sum, modern biologists find themselves in a situation highly reminiscent of the well-known joke about the man who was looking for his lost wallet under a lamppost at night. When asked whether he was sure whether this was where he had in fact lost his wallet, the man replied, "No, but this is where the light is." How can biologists be sure that the light given off by their model organisms will allow them to find the key to some important biological principle, or some mechanism common to many organisms? What grounds inferences drawn from some highly cultured laboratory organism to other, more natural organisms like humans?

From the viewpoint of the philosophy of science, this "extrapolation question" is one of the most interesting issues raised by Schaffner's chapter. Furthermore, it is (or should be) central to current debates on the Human Genome Project as well as the ethics of animal experimentation. Curiously, this question has been mostly ignored in the considerable literature on model organisms,[7] which is why Schaffner's chapter is so timely.[8]

Another issue that is more implicit in Schaffner's presentation will be more familiar to philosophers of science, but it is far from being resolved and should be re-examined in the light of current research in neuroscience: the issue of *reduction* and *reductionism* in biology. I suggest that there are important connections between the extrapolation issue and the ongoing reductionism debate. These connections are discussed in the following section. In the third section, I critically examine some objections that are likely to be raised, or have already been articulated, against Schaffner's views. In the concluding section, I draw on some recent work by Craver and Darden (this volume) and Craver (2001) on the discovery of mechanisms to shed some additional light on the role of model organisms in biological research.

Two Claims and Their Mutual Dependence

In my view, there are two significant conclusions to be drawn from Schaffner's analysis. First, recent advances in understanding the mechanisms controlling animal behavior—including social behavior—successfully use the methods and theories of contemporary genetics and molecular biology. In other words, significant progress in behavioral science is being made by isolating mutants of various species, by cloning and sequencing the corresponding genes, by deliberately introducing genetic alterations such as "knockouts," and by deploying other methods from the molecular biologist's toolkit. On the more theoretical side, animal behavior is increasingly linked to molecular processes such as signal transduction or the regulation of gene expression.

One possible way of describing these scientific developments is by saying that behavior is increasingly *reduced* to molecular mechanisms. Although Schaffner has not mentioned reduction in his chapter, I think it is fairly obvious that this is one of the major philosophical issues lurking behind it.

An important qualification to such a reductionist claim is that behavioral science is far from being *completely* "molecularized," nor is it clear that this should be its ultimate goal. As the work of Lockery and co-workers on the mechanisms underlying chemotactic behavior in the nematode *C. elegans* (Schaffner, this volume) shows, methods like modeling of neural circuits are also needed in order to secure a complete understanding of the mechanisms involved. Thus behavior is not reduced *directly* to molecules: there is at least one or possibly more than one *intermediate level* that must be understood, to a certain extent, independently of the molecular level. This is a theme that Schaffner and others have been developing for a number of years (Darden and Maull 1977; Schaffner 1993, 411–516), and the work he mentions in his chapter provides yet another example of this important insight. The main intermediate level for behavioral genetics is the nervous system—the control center for all the body movements in animals collectively known as "behavior." Thus behavior is not reduced to genetics or molecular biology alone; it is reduced to an interlevel amalgam of molecular biology, development, and neurophysiology.

The second and main thesis is that knowledge of the simple neurological systems of certain model organisms supports inferences to more complex organisms, including humans. What are the grounds for such inferences or extrapolations? (After all, there seems to be an enormous difference between the rather limited behavioral repertoire of a nematode worm and the immensely complex behavior of a primate!) As I have argued in the previous section, this question is important for all of modern biology, and it is made even more so by historians' findings on the contingencies of experimental organism choice.

The main theoretical fact licensing inferences from simple model systems to more complex organisms is that many of the mechanisms regulating development appear to be *phylogenetically conserved* among vertebrates and invertebrates. Many homologous proteins are known today that have been implicated in the same kind of process in phylogenetically distant species, such as *Drosophila*, mice, and humans. Classic examples of such proteins are the ones encoded by the homeobox genes, which all seem to be transcription factors involved in the specification of cell identity along the main body axis in a developing embryo. Another example is the *eyeless* gene, which seems to control eye development in an enormous variety of different organisms. It seems that the more "downstream" along a developmental pathway a particular mechanism is located, the harder it is for evolution to get rid of this mechanism. The same is likely to be true for some of the mechanisms controlling neurogenesis. From what we know today, it seems that evolution is extremely conservative in its deployment of molecular mechanisms. One set of these highly conserved mechanisms appears to be the ones implicated in signal transduction, that is, in the transmission of an outside signal received by a cell to the nuclear genes. Mechanisms like these play a crucial role in the development and functioning of a multicellular organism, including its nervous system.

Thus there is a considerable amount of theoretical knowledge, including knowledge about evolution, that supports extrapolations from "model organisms" to other organisms, ultimately justifying the whole notion of a model organism.

Now for the mutual dependence of reductionism and extrapolation from model organisms. The basic strategy of the (as it is sometimes called) "simple systems approach" is to try to understand some mechanism in a simple organism and then to generalize to more complex ones. If this strategy is successful, typically it will turn out that the simple mechanism is part of a more complex mechanism at some higher level, and it is assumed that it will operate similarly in this different context (Wimsatt 1998, 268). An example is the mechanism of long-term potentiation (LTP), which has been studied in very simple central nervous systems like that of *Aplysia* and is now being studied in the mouse brain. I discuss this in more detail in the concluding section. For the time being, it suffices to see that the possibility of generalizing from a simple system to a more complex one depends on the extent to which the behavior of complex systems can be explained by the properties of their parts—in other words, it depends on the extent to which reductionistic explanations are possible in the first place. Conversely, the extent to which more complex systems display *emergent properties* limits the possibility of generalizing findings from simpler systems. This is especially evident if recurrent claims about "macrodetermination" in complex systems (e.g., Sperry 1986) are correct. In this case, there is no guarantee that some component of a complex system will behave in the same way as in a simpler model system. Thus the possibility of generalizing or extrapolating from simpler systems depends on the truth of some central reductionist tenets.

Conversely the success of reductionism in biology crucially depends on the possibility of generalizing from simpler systems. What if the genetic system of humans had turned out to be radically different from, say, those of *Drosophila,* yeast, and the mouse? Clearly we would know much less about it. The fates of reductionism and of the model system approach are closely tied to each other.

It is important to appreciate the relevance of these considerations to the Human Genome Project. In particular, we are now able to understand better why so many scientists have urged that the genomes of a number of nonhuman organisms—including yeast, *C. elegans, Drosophila,* and the mouse—be sequenced along with the human genome. The project of determining the entire human genome has often been criticized—apart from the associated ethical issues—on the grounds that the sequence information would not have scientific value (if scientific knowledge is considered as a public good) commensurate with its cost (Rosenberg 1996; cf. Kitcher 1994). This might be true if only the *human* sequence became available, because then it would be extremely difficult to establish the function of large parts of the DNA sequence. However, model systems such as *C. elegans* will provide "functional annotation" (Chervitz et al. 1998) for the human genomic DNA sequence. What this means is that many DNA sequences are likely to be found in the human sequence that are homologous to sequences from a model organism, with a function that is known from studying this model organism. A striking example, again, are the mammalian *Hox* genes, which were found originally through their sequence homology to homeobox genes from *Drosophila*. Such findings are likely to occur in neuroscience as well.

Two Objections and How to Handle Them

It is now time to discuss some possible objections. The first concerns the very possibility of reductionistic genetic explanations of behavior, even in the model organisms. It could be argued that behavior is *emergent* with respect to genes. This kind of objection usually starts by pointing out that there is no simple one-to-one mapping from behavioral traits to genes. Such traits are always influenced by many genes. In addition, the effect of any gene can depend on its genetic milieu. Furthermore, the behavioral dispositions of even a simple organism arise out of complex developmental pathways involving a very large number of genes as well as other cellular constituents. Finally, the objection cites interactions of an individual with its environment as contributing to the emergence of behavior. Thus any gene's effect on behavior depends on the genetic milieu and environmental context. Cases in which a "gene for X" can be identified will be rare, and even in cases in which it is possible, not much is being explained. Or so it is argued.

Variants of this objection might appeal to developmental systems theory (DST) and criticize the overly DNA- or gene-centered view of contemporary molecular biology. DST is a recent philosophical account of ontogeny and evolution that denies any special status to DNA and to genes (e.g., Oyama 1985; Griffiths and Gray 1994).

The way to counter this kind of objection is to accept most of its premises but to deny that this makes behavior emergent in any interesting sense. The apparent emergence is at best epistemic and transient. In fact Schaffner (1998a) has shown that current molecular explanations of even the simplest behavioral mechanisms in *C. elegans* and *Drosophila* involve many genes, various complex interactions at the intermediate level, as well as environmental influences. Modern genetic explanations almost always involve a large number of causal factors, from genes to protein-DNA interactions, protein-protein and protein-membrane interactions, cell-cell interactions, and so forth. Various structures at the level of molecules, macromolecular aggregates, cells, tissues, and organs explicitly feature in such explanations. Furthermore any simple relationship between genes and traits is explicitly denied by current molecular theory. In those few cases in which a monogenetic behavioral trait is found, no biologist would claim that the identification of such a gene *alone* constitutes a sufficient explanation of this behavioral trait (although the mass media sometimes make such claims, but this is philosophically uninteresting).

If properties are called emergent just because they arise out of complex interactions, then almost everything is emergent. A nontrivial theory of emergence must show that there are properties of complex systems that are *in principle* not explainable from the properties of the parts and their interactions. Such arguments are notoriously problematic.

As far as DST is concerned, Schaffner has shown in the abovementioned work that most of the claims made by DST are actually beliefs that contemporary geneticists hold themselves. What DS theorists reject is a caricature of current genetic theory according to which genes are capable of making an organism all by themselves and without interacting with each other. The only tenet of DST that is not acceptable to most molecular biologists seems to be the one that the causal role of DNA is completely on a par with that of other cellular constituents such as membranes and maternal cytoplasm. This is known as the "parity thesis." Schaffner (1998a, 234) has sketched some arguments against the parity thesis, and I have some of my own; however, this is not the place to review them properly.[9]

It should be noted that the emergentist and developmentalist objections also affect the extrapolation issue. As we have seen, inferences from simple organisms to more complex ones rest on the high degree of *genetic* similarity among all kinds of organisms. Therefore, the less causal relevance is attributed to genes, the more precarious such inferences become.[10]

The second kind of objection is likely to spring from the environmentalist side of the nature-nurture controversy, and it concerns the extrapolation thesis. Such an objection could start by noting that the model organisms differ fundamentally from humans in that most of their behavior is *innate*, whereas most of human behavior is *culturally acquired*. Although there might be some simple learning or conditioning mechanisms in *C. elegans* and *Drosophila* and perhaps some rather more complex ones in the mouse, human behavior is characterized by an unparalleled flexibility and cannot be "encoded" in the genes. Thus nothing can be learned about human behavior by studying genes. This objection is indeed a powerful one, even if we leave the issue of free will aside. In fact, I am not sure whether this one is actually at loggerheads with Schaffner's views, since he explicitly acknowledges that there are severe limits to what can be learned about humans from model organisms.

However, I suggest that these limits might not be quite as narrow as they might seem at first sight. Human behavior, as both parties in the nature-nurture debate will agree, is strongly influenced by learning and memory. Both of these cognitive functions must have a neurophysiological basis. It could be claimed that much can be learned about these neurophysiological mechanisms through genetic studies on model organisms. To give an example, Kandel and co-workers have successfully used the marine mollusc *Aplysia* as a model organism for studying the molecular mechanisms underlying synaptic plasticity. This phenomenon is thought to play a crucial role in the human central nervous system as well. Phenomena like synaptic plasticity and the molecular mechanisms underlying them are thus likely to be *part* of any forthcoming explanation of learning and memory in humans. Despite the philosophical arguments advanced in support of functionalism, most neuroscientists today seem to think that explanations of cognitive mechanisms in the human brain cannot afford to ignore the specifics of neurons and the molecular mechanisms controlling neuron function. This is the view held by, for instance, Crick (1994).

My point is that behavioral genetics need not necessarily be committed to genetic determinism. Genes are very likely to play a role in the development and function of those neurological systems that are responsible for the very flexibility of human behavior, like learning and memory. It would be a mistake to assume that genes can only be involved in innate behavior.

Hierarchies of Mechanisms and the Role of Model Organisms

In order to achieve a better understanding of some of the issues raised by Schaffner's chapter, it will be useful to employ an explication of the concept of mechanism in biology developed by Peter Machamer, Lindley Darden, and Carl Craver (2000; see also Craver and Darden, this volume). They construe mechanisms as "collections of entities and activities organized in the production of regular changes from start or setup conditions to finish or termination conditions." The entities involved in biology are typically molecules, complexes of molecules, macromolecular aggregates such as membranes, whole cells, groups of cells, or organs. Their activities include chemical reactions and conformational changes in molecules, polarization and depolarization events in membranes, the release of signaling molecules by cells or organs, and controlled movements. Typical mechanistic explanations in biology now show how some entities and activities interact so as to produce regular changes in some biological system.[11]

Of particular interest for the present discussion is the fact that contemporary biology conceptualizes many processes in terms of a *hierarchy* of mechanisms that are involved in an organism's functioning. Craver and Darden (this volume) illustrate the hierarchical nature of mechanisms in neuroscience by using the example of spatial memory in the mouse. The neurological basis of this process is thought to be LTP in the hippocampus. LTP is characterized by lasting changes in the strength of synaptic connections. It is induced when a presynaptic neuron and a postsynaptic neuron are simultaneously active. This induction requires the activation of ion-conducting channels in the postsynaptic membrane. These ion channels are opened in the presence of the neurotransmitters glutamate and glycine. Since they are also sensitive to the drug N-methyl-D-aspartate (NMDA), they are known as NMDA receptors. One crucial mechanism that has been described explains how the NMDA receptor changes its conformation upon neurotransmitter binding. As long as the membrane is polarized, Mg^{2+} ions block the NMDA receptor's ion channel. If the membrane is depolarized by an action potential, this blockage is removed, resulting in an influx of Ca^{2+} ions. Ca^{2+} is an intracellular signal that mediates various cellular responses. In this mechanism the binding of neurotransmitter constitutes the start condition, whereas the influx of Ca^{2+} may be viewed as the termination condition. The mechanism describes how various entities (neurotransmitter, NMDA receptor, postsynaptic membrane, divalent cations) exhibit certain activities (binding, conformation change, influx, depolarization) that result in a regular series of events.

But this mechanism, considered in isolation, would not do much to explain information processing in the mouse brain. The interesting part, for the neuroscientist, is to understand how the mechanism just described is part of a mechanism operating one level above, at the level of neurons inducing LTP. A group of neurons inducing LTP, in turn, are part of a mechanism at the level of a brain structure, the hippocampus. There spatio-temporal patterns of neuronal activity are formed that correlate with the animal's location relative to objects in its environment. When the mouse forms a spatial map of its surroundings, some of these patterns of neural activity are influenced by LTP induction.

This brief sketch illustrates how neuroscientists use hierarchies of mechanisms to explain an example of information processing in the brain. Craver (2001) identifies four levels in this hierarchy: (1) the level of the NMDA receptor mechanism, (2) the level of single neurons inducing LTP, (3) the level of the hippocampus forming certain patterns of neuronal activity, and (4) the level of the whole organism—the mouse trying to orient in a maze. At each of these levels—except (4)— the main mechanism constitutes a *stage* in the next higher-level mechanism. For instance, the opening of an NMDA receptor complex is described by a mechanism at level (1). This event constitutes a stage in a mechanism at level (2)—the induction of LTP in an individual synapse. The formation of LTP, in turn, is a stage in the mechanism of spatial map formation in the hippocampus, which involves complex neural activation patterns influenced by LTP. Finally, spatial map formation is a stage in the mechanism at the highest level—the mouse learning to orient in the maze.

Note how such a neurobiological explanation has gaps: scientists are interested only in certain salient features of these mechanisms, namely those that are stages in some higher-level mechanism. Furthermore, as Machamer et al. (2000) point out, the hierarchy of mechanisms "bottoms out" and "tops off." In other words, neuroscientists do not care what happens on the levels below the NMDA receptor mechanism or above the level of behavior. Clearly it is a historically contingent matter at what levels the mechanistic explanations bottom out and top off, respectively. Nevertheless, these points may be part of the very definition of a scientific discipline.

I suggest that the typical explanatory structure of contemporary biology, as aptly captured in Craver's work (see also Craver and Darden, this volume), should be taken into account when grappling with the kinds of issues raised by Schaffner and also in discussions of the Human Genome Project.

First, we are now in a better position to understand why extrapolation from model organisms can work: *lower-level mechanisms may be the same in many organisms and can thus be studied in a suitable model system.* This will typically be the case for molecular mechanisms like the neurotransmitter-gated opening of the NMDA receptor.[12] Although the identity of the neurotransmitter may vary, and the receptor-ion channel complex may be embedded in a different higher-level mechanism, the basic activities involved in the process are the same. Other well-known examples of widespread molecular mechanisms are those associated with G proteins[13] involved in signal transduction, which are thought to function in a great variety of signaling processes, including processes that, when they malfunction, can cause malignant cell growth. G proteins and the associated signaling mechanisms are studied in various model organisms, including *C. elegans* and *S. cerevisae.* The extent to which such low-level mechanisms are conserved in evolution is remarkable.

The hierarchical structure of biological explanations, together with the empirical fact that many low-level mechanisms are conserved, means that mechanisms described in some model organism can be transferred (with the necessary precautions) into an explanatory structure for some other process in a different organism, where it will perform the same explanatory work. Biology today is in possession of a growing store of entities and mechanisms from which explanatory accounts of various biological processes can be assembled (cf. Craver and Darden, this volume, 123–24). Of course, verifying that some mechanism studied in a model organism performs the same activity in a different process in a different organism may involve painstaking research.

But what about higher-level processes? Schaffner seems to suggest that extrapolations may be possible even for mechanisms at the level of neural circuits or behavior. It is at this point that I sound a cautionary note. Although I do not want to exclude the possibility that crucial insights—for example, into the neural circuits of the vertebrate brain—may come from worm studies, I think that there are clear limits to extrapolation. I illustrate this with reference to the example of social behavior in C. *elegans* discussed in Schaffner's chapter: that of the *npr-1* gene, which is thought to provide a genetic explanation for social behavior of the worm. In this example, a single nucleotide substitution can make the difference between solitary or social feeding behavior. Schaffner has explained the special sense in which it could be said that this nucleotide substitution is the *cause* of the difference in the behavioral trait, and I have no quarrel with this. I would not even object to calling *npr-1* a gene "for" social behavior in C. *elegans*. However, it seems to me that extrapolation to more complex animals or to humans is not warranted in this case. Although *npr-1* codes for a putative neuropeptide receptor that is also found in the human brain in a similar form (neuropeptide Y), it is far from clear that it is embedded in the same or even a similar kind of higher-level mechanism in both cases. In particular, it would be naïve to think that the human homologue of *npr-1* is also involved in human feeding behavior or even any other kind of social behavior.[14]

Again, the notion of a hierarchy of mechanisms allows us to see more generally when and where extrapolation will work and where it might go wrong. Many higher-level mechanisms in multicellular organisms will be composed from the basic toolkit of evolutionarily conserved cellular and molecular mechanisms. We have seen how this is reflected in the way in which biologists can assemble explanatory accounts of higher-level mechanisms by using lower-level mechanisms known from the various model organisms. However, it is always possible that in some organism(s) of particular interest the lower-level mechanisms are employed differently in higher-level mechanisms. The case of *npr-1* is a likely example. Of course, there may be surprises like the *eyeless* story, in which a very similar transcription factor seems to regulate eye development in taxonomically extremely distant animals.

But biologists can never really be sure whether some such mechanism is part of a similar kind of higher-level mechanism until they carry out a suitable *investigation*.

This brings me to a final, also somewhat critical remark. The question is whether Schaffner's talk of "extrapolation" or "inference" gives us an adequate description of the use of model organisms in contemporary biology. The term "extrapolation," as it is normally used in science, refers to the supposition of a mathematical function in a region of the variables' domain where no values are known. "Inference" refers to the assertion of statements that, by virtue of their logical form, are entailed (deductive inference) or made more probable (inductive inference) by some other statements. Generally we are talking here of *logical* and *evidential relations*. Is this really the correct picture in the present context? Do biologists consider knowledge of mechanisms obtained in a model organism as *evidence* that, given the necessary evolutionary relationships, a similar process operates in other organisms? I think that such a picture would be misleading. For whether or not a certain higher-level process is implemented by certain lower-level processes in a given organism is still a question to be answered by *empirical research*. In other words, a mechanistic account of some process in a model organism is not *evidence* for any specific hypothesis on the mechanisms involved in other organisms, even if it is combined with knowledge about the evolution of these mechanisms.[15] It is always possible that evolution has appropriated a given mechanism for some other function.

I suggest that knowledge of mechanisms operating in model organisms plays a different role in research than providing a basis for extrapolations or premises for some sort of inductive inference. First, such knowledge performs an important *heuristic* function in guiding research; it tells biologists where to look and what to look for. Second, work on model organisms may provide important *research tools* for studying other organisms—including DNA sequence information for searching for homologies, DNA probes for cloning, and antibodies. Third, studies of model organisms are necessary for building and completing the store of mechanisms and entities for assembling complex higher-level mechanisms out of basic lower-level mechanisms and entities. I do not think that the notions of "extrapolation" or "inference from model organisms" capture these three crucially important roles of model organisms in contemporary biological research.

Summary and Conclusion

I have tried to place the issues raised by Schaffner's chapter within a broader context of recent historical and philosophical studies of biological research. Clearly the issue of extrapolating from model organisms is relevant not just to neuroscience but also to most areas of experimental biology. I have shown that the issue of "extrapolating" from model organisms can be brought to bear on some of the challenges raised by those historians of biology who have studied experimental practices and who were driven to the conclusion that noncognitive, contingent factors affect the choice of model organisms and therefore the whole research process—possibly including the kinds of models or theories that will be established. The upshot of this discussion is that for the validity of general or middle-range theories about biological processes, it does not seem to matter much which model organisms scientists choose, as long as knowledge of a particular organism can be placed within the proper comparative evolutionary context. Furthermore, I have pointed out some connections between the "extrapolation issue" and another long-standing debate in the philosophy of science, namely reduction and reductionism, and I have examined some possible emergentist[16] and developmentalist objections to the view that reductionistic explanations for neurological processes and behavior can be found with the help of research on model organisms. I have applied the general framework developed by Craver and Darden (this volume) and Machamer et al. (2000) on the discovery of mechanisms to explicate further the ways in which model organisms function in biological research. We have seen that much current biological research functions in a peculiar way: biologists produce explanatory accounts of biological phenomena by assembling hierarchies of mechanisms from a toolkit of basic lower-level mechanisms that are evolutionarily conserved. Model organisms provide access to this toolkit. Finally, I have argued that the terms "inference" and "extrapolation" should not be taken too literally in this context.

NOTES

I thank Ken Schaffner for stimulating discussion, Carl Craver for making unpublished work available to me, and Eric Oberheim for critically reading the manuscript.

1. I shall argue that the notions of "extrapolation" and "inference" are somewhat misleading in this context. But in order to develop the issues, I shall stick to these notions for the time being.

2. A classic example of this kind of historical analysis is Kohler's (1994) detailed account of the development of *Drosophila* as a model organism for genetics. See also Burian (1992, 1993), Mitman and Fausto-Sterling (1992), Clause (1993), Holmes (1993), Kohler (1993), Lederman and Burian (1993), Lederman and Tolin (1993), Summers (1993), Zallen (1993), de Chadarevian (1998), Geison and Creager (1999), and Rader (1999).

3. The most extreme form of such a claim is encapsulated in Jacques Monod's famous dictum "What's true for *E. coli* is true for the elephant," which turned out to be false in ways not foreseeable by Monod. However, a weaker form of this belief seems to prevail among most molecular biologists.

4. A possible example is Wilhelm Roux's "mosaic theory" of development, which he might not have developed had he not been working with frog embryos, but instead with Driesch's "regulative" embryonic systems (Weber 1999 and references therein).

5. A classic example is *Drosophila*. In the typical laboratory strains, males are deficient for homologous recombination, which greatly facilitates the interpretation of crossing experiments.

6. However, claims that laboratory organisms are "constructed," "invented," or "created" (e.g., Rader 1999) are exaggerated, as they ignore the still largely biological origin of laboratory organisms.

7. An exception is Burian (1993, 365), who asks: "To what extent are the results obtained with an organism (or a group of organisms) general, and to what extent can they be reliably extrapolated?" Among his most important conclusions, in my view, are (1) that biology deals with "contingently different systems whose regularities, even if they trace back to fundamental (e.g., biochemical) laws, depend at least as strongly on the contingent, evolutionarily derived configurations of the components of the system as they do on those laws"; (2) that "the epistemological situation of biology is different from that of any form of mechanics, including quantum mechanics, for *biological knowledge is knowledge of a large number of particular systems that cannot be identically prepared*"; and (3) that "proper evaluation of the knowledge gained by working with a given organism or a group of organisms requires that knowledge to be set into a comparative and evolutionary framework" (366, emphasis in original). It is precisely for these reasons that I disagree with Burian's claim that we have here "an especially acute version of the traditional philosophical problem of induction" (366), especially if the problem of induction is understood along the lines of Goodman's (1973) "new riddle of induction." As I elaborate later, the problem of generalizing from model organisms is not a matter of inductive inference, but one of comparative empirical research. Inductive inference is only relevant insofar as it underlies all empirical research (which is controversial).

8. See also Schaffner's (1998a) important paper and the discussions by Gilbert and Jorgensen (1998), Griffiths and Knight (1998), Schaffner (1998b), and Wimsatt (1998).

9. Briefly, I reject the parity thesis on the grounds of the enormous phenotypic differences that differences in DNA sequence can cause, where "cause" is understood in Mackie's sense of necessity in the circumstances (Schaffner, this volume). After all, it is the difference in DNA sequences that makes the considerable phenotypic difference between, say, a human and a gorilla; the maternal cytoplasms are extremely similar. No other constituent of a developing organism has that kind of effect on the range of possible phenotypes.

10. Furthermore, Gilbert and Jorgensen (1998, 259) hint that there are reasons for the developmentalist to be suspicious of the entire model system approach: "The very richness of life that the Developmentalist Challenge claims engenders diversity has been hunted down and eliminated from *C. elegans* research."

11. Note that many biological processes—for example, muscle contraction, some metabolic pathways, and DNA replication (my examples)—are *cyclical,* such that setup conditions and termination conditions are identical.

12. Similarity of mechanisms might also occur at the level of neural circuits, as Schaffner (this volume) suggests.

13. G proteins are found in a variety of regulatory and signal transduction mechanisms. They are characterized by a guanosine triphosphate binding site that regulates the enzyme's activity.

14. Another recent example is the *lov-1* gene in *C. elegans,* which is implicated in male mating behavior. The closest human homologues for *lov-1* are *PKD-1* and *PKD-2,* which are kidney disease loci (Barr and Sternberg 1999). As such, they are unlikely to be involved in human male mating behavior.

15. Again, the *C. elegans npr-1* gene discussed by Schaffner may serve as an illustration.

16. In the discussion that followed presentation of his paper at the conference on which this volume is based, it became clear that Schaffner is more of an emergentist than I am. However, it would be necessary to clarify our concepts of emergence before we could really argue this point. My skepticism concerning emergence concerns a *strong* sense of emergence, that is, one that claims the impossibility of a reductionistic explanation for some *ontological* reason like macrodetermination.

REFERENCES

Barr, M. M., and W. Sternberg. 1999. "A Polycystic Kidney-Disease Gene Homologue Required for Male Mating Behavior in *C. elegans.*" *Nature* 401: 386–89.

Burian, R. M. 1992. "How the Choice of Experimental Organism Matters: Biological Practices and Discipline Boundaries." *Synthese* 92: 151–66.

———. 1993. "How the Choice of Experimental Organism Matters: Epistemological Reflections on an Aspect of Biological Practice." *Journal of the History of Biology* 26: 351–67.

Chervitz, S. A., L. Aravind, G. Sherlock, C. A. Ball, E. V. Koonin, S. S. Dwight, M. A. Harris, K. Dolinski, S. Mohr, T. Smith, S. Weng, J. M. Cherry, and

D. Botstein. 1998. "Comparison of the Complete Protein Sets of Worm and Yeast: Orthology and Divergence." *Science* 282: 2022–28.

Clarke, A. E., and J. H. Fujimura, eds. 1992. *The Right Tools for the Job: At Work in Twentieth-Century Life Sciences*. Princeton, N.J.: Princeton University Press.

Clause, B. T. 1993. "The Wistar Rat as a Right Choice: Establishing Mammalian Standards and the Ideal of a Standardized Mammal." *Journal of the History of Biology* 26: 329–49.

Craver, C. F. 2001. "Role Functions, Mechanisms, and Hierarchy." *Philosophy of Science*, March.

Crick, F. 1994. *The Astonishing Hypothesis*. London: Simon and Schuster.

Darden, L., and N. Maull. 1977. "Interfield Theories." *Philosophy of Science* 44: 43–63.

De Chadarevian, S. 1998. "Of Worms and Programmes: *Caenorhabditis elegans* and the Study of Development." *Studies in History and Philosophy of Biological and Biomedical Sciences* 29C: 81–105.

Geison, G. L., and A. N. H. Creager. 1999. "Introduction: Research Materials and Model Organisms in the Biological and Biomedical Sciences." *Studies in History and Philosophy of Biological and Biomedical Sciences* 30C: 315–18.

Gilbert, S. F., and E. M. Jorgensen. 1998. "Wormholes: A Commentary on K. F. Schaffner's 'Genes, Behavior, and Developmental Emergentism.'" *Philosophy of Science* 65: 259–66.

Goodman, N. 1973. *Fact, Fiction and Forecast*. Indianapolis: Bobbs-Merrill.

Griffiths, E., and R. D. Gray. 1994. "Developmental Systems and Evolutionary Explanation." *Journal of Philosophy* 91: 277–304.

Griffiths, E., and R. D. Knight. 1998. "What Is the Developmentalist Challenge?" *Philosophy of Science* 65: 253–58.

Hacking, I. 1983. *Representing and Intervening: Introductory Topics in the Philosophy of Natural Science*. Cambridge: Cambridge University Press.

———. 1992. "The Self-Vindication of the Laboratory Sciences." In A. Pickering, ed., *Science as Practice and Culture*. Chicago: University of Chicago Press, 29–64.

Holmes, F. L. 1993. "The Old Martyr of Science: The Frog in Experimental Physiology." *Journal of the History of Biology* 26: 311–28.

Kitcher, P. 1994. "Who's Afraid of the Human Genome Project?" In D. Hull, M. Forbes, and R. M. Burian, eds., *PSA 1994*. East Lansing, Mich.: Philosophy of Science Association, 313–21.

Kohler, R. E. 1993. "*Drosophila*: A Life in the Laboratory." *Journal of the History of Biology* 26: 281–310.

———. 1994. *Lords of the Fly:* Drosophila *Genetics and the Experimental Life*. Chicago: University of Chicago Press.

Lederman, M., and R. M. Burian. 1993. "The Right Organism for the Job: Introduction to Special Section." *Journal of the History of Biology* 26: 235–37.

Lederman, M., and S. A. Tolin. 1993. "OVATOOMB: Other Viruses and the Origins of Molecular Biology." *Journal of the History of Biology* 26: 239–54.

Machamer, P., L. Darden, and C. Craver. 2000. "Thinking about Mechanisms." *Philosophy of Science* 67: 1–25.

Mitman, G., and A. Fausto-Sterling. 1992. "Whatever Happened to Planaria? C. M. Child and the Physiology of Inheritance." In Clarke and Fujimura 1992, 172–97.

Oyama, S. 1985. *The Ontogeny of Information: Developmental Systems and Evolution.* Cambridge: Cambridge University Press.

Rader, K. A. 1999. "Of Mice, Medicine, and Genetics: C. C. Little's Creation of the Inbred Laboratory Mouse." *Studies in History and Philosophy of Biological and Biomedical Science* 30C: 319–44.

Rheinberger, H.-J. 1997. *Toward a History of Epistemic Things: Synthesizing Proteins in the Test Tube.* Stanford, Calif.: Stanford University Press.

Rosenberg, A. 1996. "The Human Genome Project: Research Tactics and Economic Strategies." *Social Philosophy and Policy* 13: 1–17.

Schaffner, K. F. 1993. *Discovery and Explanation in Biology and Medicine.* Chicago: University of Chicago Press.

———. 1998a. "Genes, Behavior, and Developmental Emergentism: One Process, Indivisible?" *Philosophy of Science* 65: 209–52.

———. 1998b. "Model Organisms and Behavioral Genetics: A Rejoinder." *Philosophy of Science* 65: 276–88.

Sperry, R. W. 1986. "Discussion: Macro- versus Micro-Determinism." *Philosophy of Science* 53: 265–70.

Summers, W. C. 1993. "How Bacteriophage Came to Be Used by the Phage Group." *Journal of the History of Biology* 26: 255–67.

Weber, M. 1999. "Hans Drieschs Argumente für den Vitalismus." *Philosophia Naturalis* 36: 265–95.

Wimsatt, W. C. 1998. "Simple Systems and Phylogenetic Diversity." *Philosophy of Science* 65: 267–75.

Zallen, D. T. 1993. "The 'Light' Organism for the Job: Green Algae and Photosynthesis Research." *Journal of the History of Biology* 26: 269–79.

12

Consciousness

A Neurological Perspective

Paul-Walter Schoenle

Lurija Institute for Rehabilitation Sciences and Health Research, University of Konstanz, and Kliniken-Schmieder, Allensbach, Germany

It is not possible to give a complete and coherent outline of a neurological theory of consciousness in this chapter. Rather the goal is to point out some of the rich information that neurology and neurological rehabilitation can contribute in the general search for a theory of consciousness.[1] However, the neurology of consciousness also challenges whatever theory of consciousness there might be, in that such a theory, for theoretical reasons,[2] must be able to account not only for normal consciousness (the physiology of consciousness) but also for alterations of consciousness (the pathophysiology of consciousness)—which may arise from brain alterations due to diseases of the brain, trauma, toxins, or drugs—and, last but not least, for the institutional restorative processes underlying the recovery of impaired consciousness.

Neurology and neurological rehabilitation deal with a variety of alterations in consciousness ranging from coma to blind sight and confabulation (table 12.1). We can observe these phenomena more often today because more patients than ever survive even severe brain damage resulting from accidents, violence, brain bleeding (hemorrhages), ischemic brain infarcts, or cerebral hypoxia during heart failure. Survival has become possible because of advances in drugs and medical techniques in recent decades, including emergency and resuscitation medicine, surgery (especially neurosurgery), and intensive care medicine. Most patients, however, not only survive but also re-

TABLE 12.1 Alterations of Consciousness in Neurological Disorders

Coma (cerebrovascular accidents; traumatic brain injury; encephalitis; tumor; toxic, metabolic, endocrine disturbances)	Asomatagnosia
	Anosognosia
	Simultanagnosia
	Alien hand syndrome
Vegetative state (apallic syndrome)	Capgras syndrome
Akinetic mutism	Fregoli syndrome
Low consciousness state	Delusional reduplication of self
Hypersomnia	Two-life syndrome
Abulia	Loss of self-identity (self-awareness, "disautia")
Absence and partial complex seizures	
Acute confusional state	Confusional state
Confabulation	Confabulation
Hemineglect	
Blind sight	
"Blind" touch	

cover to a better life—thanks to intensive care rehabilitation (ICR), the newest branch in medical rehabilitation. It helps ensure rehabilitation immediately after definitive acute intervention, even if such intensive care measures as artificial ventilation are still required.[3] ICR units provide the unique opportunity to observe large numbers of patients in transient or prolonged comatose or comalike states and in other states of profound functional impairment over extended periods of time and to observe transitional states in those emerging from coma to normal consciousness. Often the conditions exhibited by such patients fall under no clear syndrome labels or else have only such tentative names as "low awareness state" or "minimally reactive, minimally conscious" state. The re-genesis of mind with the emergence from coma and the re-establishment of mental functions have become the clinical and scientific leitmotiv of ICR.

Consciousness for Action: Neurorehabilitative Aspects

The goal of neurorehabilitation (NR) is to enable a patient to become independent again in activities of daily living and to participate in his social and vocational environment. This requires the ability to be aware of the self and the environment in order to perform appropriate actions at the right time. In this respect—that the primary goal is to re-establish the patient's ability to act—NR's conceptual framework is in part based on a pragmatic, teleological view on the brain: the human being and the brain are an action and abstraction machine.[4] The brain is the pragmatic control system, designed to ensure that the host survives in his environment.[5] In order to optimize its control function the brain is also an abstraction device. It allows abstraction of a representational model of the world from the outer concrete, presentational[6] environment and, simultaneously, the construction of a representation of the self embedded in the world model[7] that constitutes the awareness of the self and its environment. (This is the representational or phenomenological hypothesis of consciousness.) The abstract representations can be manipulated and possible world models computed, as projected onto the present environment.[8] Possible effects of the self's intervention in the environment can be anticipated, and plans can be evaluated in order to select the best actions to be undertaken (strategic thinking and planning), including the control of the effects of the actions (for further details see Schoenle in press). In the representational virtual world, programmed in its own user code,[9] the self can directly view the world and act on it without being hampered by the physical properties of its own machinery and those of the environment. When necessary, the self can dynamically expand itself, incorporate tools and instruments (e.g., a car, a knife, a pen) as part of the body's representations, and use them as automatically and integrally as its own body parts. Optimal action planning involves permanent updating[10] of information about the self and the environment and permanent readiness for action. Most of these processes must be performed automatically in order not to flood the control system with sensory information and overload it with cognitive and motor processing during the planning of motor actions. The control system takes over only when new operations must be performed, for example, when novel information comes up or new motor activities (skills) must be learned. It establishes processing routines and transfers them to specifically dedicated modules that can act automatically. As many processes as possible are automatized, be they the identification of known faces or objects, the production of sentences, or the execution of "prefabricated" skilled movements.

As a principle, the brain tends to operate in an automatic mode, to set itself free to cope with novel situations and produce new solutions that require controlled processing and demand large amounts of attentional resources. These special tasks are performed by what can be called a general-purpose processing system (GPPS). In cases of highly automatized processing no increase in local metabolism or perfusion can be observed in functional magnetic resonance imaging (fMRI) or positron emission tomography (PET) studies, whereas in controlled processing situations local neuronal activity increases, as do metabolism and circulation, in the reticular system, prefrontal areas, basal forebrain, and the other limbic structures that cooperate in learning and memory formation (e.g., the hippocampus) and make up the GPPS. When controlled processing is required, high amounts of glucose and oxygen are utilized owing to an increase in the level of enzymatically based intracellular processes, including induction of protein synthesis.[11]

Loss of *automatization* is a common consequence of brain damage. It leaves patients with painstaking, psychic energy–consuming, and often frustrating attempts to consciously re-establish formerly automatic (i.e., unconscious) actions and behaviors, such as producing a common word or even a simple phoneme sequence. Some patients, especially after traumatic brain injury, suffer when high demands are placed on conscious controlled processing involving selective attention. Some (most often those with frontal lobe damage) have lost the ability to operate as an open, *self-aware, self-learning, and self-organizing* system that can cope with novel situations (e.g., going on vacation or beginning a new job). These patients function quite normally and behave appropriately if external control and structure are provided. Yet without external management they are lost and would not survive.

From a practical NR and an evolutionary perspective, *consciousness can be viewed as the most important and most general human ability to act as an open, self-aware, self-learning, and self-organizing system for adaptation to novel situations.* It controls information acquisition, abstraction for representation of self and environment (self and environmental awareness),[12] manipulation of these representations (modeling of self and environment and future states), and action, including planning and execution. *Consciousness in this sense is the action control system for survival.*

Neurofunctionalism

At the onset of a neurological disorder the patient or her relatives have the experience that something is not functioning as usual ("I cannot move my arm." "Lately she always gets lost." "She slipped out of the chair and was unconscious"). A clinical neurological examination is performed to determine the functional disturbances and to narrow them down to a point at which hypotheses about the underlying etiology can be formulated and, one hopes, a definitive diagnosis can be made. In many cases complementary technical investigations are required.[13]

The clinical examination is based on a function-analytic, neurofunctional[14] approach that, in acute neurology,[15] is used only in a narrow, target-oriented way to identify the underlying structural alteration in the nervous system, to determine its etiology, and to develop a treatment strategy. In NR the approach is more detailed. It is based on a comprehensive analysis of a particular function, which is conceived of as part of a functional system (e.g., grasping, naming an object, comprehending a sentence, recognizing a face). The analysis involves decomposition into functional components and an understanding of the operational mode of the system.[16] The goals of the function-analytic approach are, first, to identify the component causing the pathological output of the system; second, to correlate it with the neuropathological substrate (lesion localization), including the underlying disease process; and, third, based on that information, to re-establish the overall functional output of the system.

Thus when treating patients with impaired consciousness the clinical approach is based on detailed analysis of both the behavioral and the underlying structural levels. The results are then mapped onto the clinician's functional and structural knowledge about consciousness and its disorders. In the clinical approach it is implicitly assumed that consciousness is a multiple-component, complex functional *and* structural system. It is further assumed that impairment of consciousness varies with the neuronal components damaged, that is, that different types of brain damage result in different types of impaired consciousness (the multiple-consciousness hypothesis). Some evidence in support of this assumption is provided by its usefulness and efficacy in treating comatose patients.

In a simplified view of the structural level, the functional components (table 12.2) appear to be implemented in two principal neuroanatomical systems.[17] The first system, the primary consciousness system (PCS), is organized along the neuraxis. It emanates from the spinal cord and runs up to the frontal lobe. It is the basic system,

TABLE 12.2 Major Functional Components of Consciousness

Alertness/arousal/wakefulness	Sleep/wake cycle
Drive, motivation	
Orienting and vigilance	Providing information
Sensation and perception	Providing information
Attention	Focusing, filtering, selection
Memory	Recovery, display, and spatial and temporal ordering of information; relation of remembered events in time and space
Working memory/intention/executive/ action component	Related to attention and important for various processes, such as continuously presenting information, integration from past and present experience for making minute-to-minute decisions, providing self-awareness, providing awareness of some of one's own cognitive and mental processes, self-reference, checking of information for its relevance, interpretation of the present, "conscious" experience and awareness of the environment, anticipating experience, thinking, deductive reasoning, planning, and decision-making

having to do with arousal, alertness, drive, and attention, including working memory and executive functions.[18] The primary system is necessary in that its operation is a precondition for all other (especially cognitive) functions involved in consciousness. However, it is not sufficient; a second system is needed that contributes various kinds of information (e.g., external and internal environmental sensory and perceptual information or affective and emotional information)—the qualia that form the various qualities of consciousness. This secondary system forms the dominating type of consciousness prevailing at a given moment: "pain consciousness" dominates when pain, from a dental abscess or stepping on a nail, is of highest importance to the subject; "emotional consciousness" when feelings of love are dominant; "perceptual (auditory)" consciousness when one is listening to a concert; or "action consciousness," when one is performing complicated tasks (figure 12.1). The secondary consciousness system (SCS) is formed by the various cortical areas, especially sensory and motor primary and association areas, as well as structures of the limbic sys-

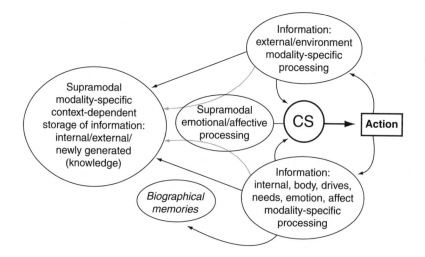

CS

Evaluation/selection
Integration/coherence of information
"Content"
Generation of new information (modeling)
Abstraction
Anticipation
Action selection/planning
Execution control supervision

Figure 12.1 Processing model of consciousness as an action-subserving control system with continuously updated experience into awareness (CS) of self and environment as an undulating (tonic, phasic) activity that is ongoing throughout life. Information about the body, the internal situation (including drives, needs, emotions, and affect), and the environment is picked up and processed in a modality-specific way and then integrated supramodally with affective coloring before it enters conscious experience on its way to long-term storage of experienced (biographical) memories. The focus of CS can shift more to internal, external present, or past experiences. It can narrow or expand or become fractionated. It can also become disintegrated, as is the case in various brain dysfunctions. Processes related to the CS are listed at the lower left.

tem. The multiplicity of consciousness arises from the dynamic association (interaction) of the two systems and their components, the PCS being more of a general-purpose processor and the SCS being more of a dedicated processing system. The latter is specialized for the processing of distinct information (e.g., in the visual domain) in an automatic way and supplies the primary system with informational content that emanates as conscious qualia.

The distributed multicomponent structure of both systems has several advantages as compared with an architecture in which all functions of consciousness are encapsulated within a single module.[19] First, it is economical in that specific processing modules can be applied to various tasks and be recruited into work groups from moment to moment whenever needed.[20] Multiple consciousness can be instantiated by recruitment of varying components of the second system. The second advantage is that when brain damage occurs the resulting impairment is more likely to be limited; some components remain intact and can reorganize the functional system with degraded functionality or by relying on recruitment of other areas. Thus the system can keep the individual "somewhat" conscious or alive and, if the impairment of the affected components was only transient, the whole system can recover.

Several gross predictions can be made about consciousness impairments if brain damage occurs in such a way that it affects the two systems distinctively. First, complete dissociation of the two systems should result in wakeful consciousness without access to information that is still available for unconscious processing (blind sight, hemineglect, and many others, including some vegetative states); the information is locally intact but cannot be transformed into conscious qualia. Second, impairment of the secondary system alone should have the same result, in that patients are awake but unable to form conscious qualia owing to a loss of local qualia information; in addition, unconscious processing should be impaired as well. Third, impairment of the primary system should result in a complete breakdown of consciousness as there is no arousal or awareness, so the qualia cannot be processed into consciousness (i.e., the patient is in a coma). If components within the primary system are damaged differentially (e.g., along the arousal-activation line), then a loss of the arousal component should result in coma with unavailability of the awareness component. When the arousal component recovers, the patient may be in an awake conscious state, but one in which the information about the qualia are intact but cannot be brought to awareness. Unaware, implicit processing should be possible. If there is not complete loss but only dysfunction of awareness, then the information that would constitute the qualia cannot be integrated coherently and confusion and confabulation will result—qualia do not occur and there is no consciousness of them.

Neuroanatomical Aspects and Neurofunctional Implications

In this section an outline of the neuroanatomical structures is given as they relate to consciousness. The PCS serves for arousal and activation and for attention and executive functions; it is oriented along the neuraxis. The SCS is mainly cortical and provides specific information to form qualia representations of self and environment.

The Primary Consciousness System

The Reticular Formation at the Pontine and Midbrain Brain Stem Level. The *reticular formation* (RF) plays a central role in control of arousal/alertness and attention.[21] It forms an important part of the brain stem (figures 12.2 and 12.3) and comprises three sets of neural nuclei—the median, medial, and lateral—in the core and dorsal parts of the brain stem, stretched out in a rostrocaudal direction and running from the anterior midbrain to the pons and the medulla oblongata with a continuous transition into the central gray matter of the spinal cord. Rostrally the RF extends to the intralaminar nuclei and the reticular nucleus of the thalamus, and to the hypothalamus, the basal forebrain, and the anterior cingulate cortex (AC) (Nieuwenhuys et al. 1991).

The medial RF (MLRF) receives input from all sensory modalities and the cerebellum, and it forms the *ascending reticular activating system* (ARAS) with direct and indirect projections to the intralaminar thalamic nuclei (ILN), which in turn project to the striatum and, in topographical order, to cortical areas. *The MLRF controls the sleep-wake cycles and the activity level of the cortex.* The midbrain RF neurons (mostly cholinergic) connect via a dorsal branch to the ILN and via a ventral branch to the reticular nucleus thalami, the subthalamic nuclei, and the hypothalamus.

The median RF (MNRF) consists of the raphe nuclei, which project *serotonergically* to those components of the limbic system that are engaged in *cognitive processing*, such as the hippocampus, amygdala, basal forebrain, limbic nuclei of the thalamus, and cingular and entorhinal cortex, as well as to the frontal, temporal, and parietal cortex.

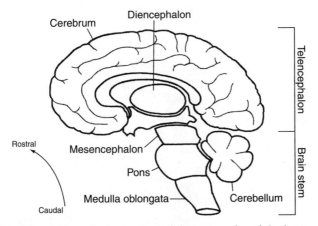

Figure 12.2 Neuroanatomical overview of the geography of the brain and its principal nomenclature.

The lateral RF (LRF) is mainly formed by the *noradrenergic* locus coeruleus (LC), which also connects to all *cognitively relevant limbic components* and plays a major role in control of *attention and the supervision and surveillance of ongoing external events and internal, autonomic states.* LC has regulatory influence over the hypothalamus and copes with *emergency situations.* LC's connection to the prefrontal cortex (PFC) can communicate the importance and relevance of complex sensory events and situations.[22] The neurons and fibers of the LRF also relate to the autonomic nervous system and *autonomic cardiovascular* and *respiratory* functions; they are located in the lateral tegmental parts of the RF.

In general, the medial and lateral RF plays more of an activating role, and the median RF more of an inhibitory role. In cooperation they develop a strong capability for focusing on certain sensations and perceptions.

The RF is the most basic and life-sustaining integration and control system of the central nervous system. It not only is the crossroad for all afferent and efferent projections[23] but also contains the autonomic control centers for cardiac, respiratory, and masticatory functions, including those of the cranial nerve nuclei,[24] and has a profound influence on arousal and the activity level of the cortex. In addition, it acts as a relay for the transmission between the periphery of the body and

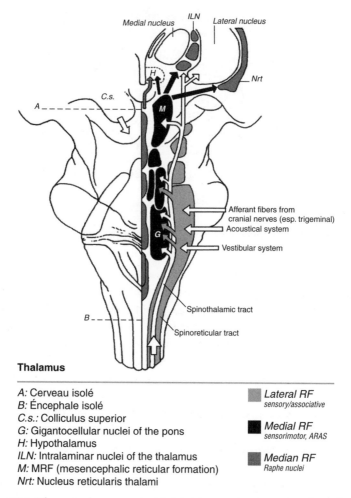

Figure 12.3 The reticular system, with its three portions running through the brain stem from the medulla oblongata through the midbrain (mesencephalon) to the thalamic structures of the ILN and NRT. (Modified from Zschocke 1995, 31.)

the cerebral hemispheres of information necessary for perception formation.

The Thalamic Level. The midbrain reticular formation (MRF; figures 12.3 and 12.4) has major projections to the intralaminar nuclei of the *thalamus* (ILN)[25] and to the thalamic reticular nucleus (NRT), including the lateral thalamic nuclei and the zona incerta.[26] The ILN and NRT can be viewed as the rostral extension of the midbrain and pontine ARAS. They play an important role in the control of consciousness by gating sensory information flow to cerebral cortex, achieving polysensory integration, and linking widely distributed cortical areas for percept formation.

The ILN receive sensory information through the spinal cord (spinothalamic tract) and input from many cortical areas, which they project back in topographical order, preferentially to the PFC. In addition, they are heavily interconnected with the limbic system. They function as a relay of the ARAS for cortical arousal, act as a thalamic pacemaker, and demonstrate stimulus-dependent synchronization with cortical cells. They perform polysensory (visual, auditory, and somesthetic) integration and sensorimotor integration. As part of the motor system they modulate the input and output of the striatum (via parafascicular and centromedial nuclei) and inhibit and release motor plans and actions (trigger function), including the control of eye movements (important in explorative, attentional behavior). Finally, they play an important role as central pain modulators.

The NRT is crucial for *gating sensory information flow to the cerebral cortex*. It envelops most of the ipsilateral thalamus, allowing thalamocortical and corticothalamic fibers to pass through, and exerts a tonic inhibition on the adjacent thalamic relay nuclei through its GABAergic (inhibitory) efferents. When activated by MRF, NRT's inhibitory power is reduced (disinhibition) to allow sensory transformation to be transmitted to cortical areas. As MRF's activation of NRT is topographically organized, inhibition and disinhibition occur in a selective and focal way. This gating of sensory influx into the cerebral cortex is also controlled by input that NRT receives from PFC, allowing certain stimuli to be selected or rejected and for the attenuation of sensory input according to its novelty and relevance, thus serving as a mechanism for selective and focused attention.

NRT also plays an important role in linking widely distributed cortical areas engaged in the processing of certain sensory stimuli. Based on its intrinsic oscillatory activity (20–40 Hz), it drives the corticothalamic re-entrant system under the modulatory influence of MRF. Stimulation of MRF facilitates synchronization of gamma discharges (30 Hz) among spatially separated cortical neuronal pools in response to sensory stimuli. In this way, various aspects of a stimulus can be processed separately in distant cortical modules, with the coherence of the stimulus preserved by the phase-locked neuronal activity (Munck 1996).

The importance of MRF and ILN for arousal and vigilance as well as for the transition between sleep and wake states has been demonstrated by a PET study (Kinomura et al. 1996). Both the midbrain reticular formation and the thalamic intralaminar nuclei became activated when human participants went from a relaxed waking state to an attention-demanding reaction-time task. These structures are engaged in maintaining a state of high *vigilance and attention* independent of sensory stimuli and motor reaction.

For the *integration of consciousness with the action system* some other nuclear components of the thalamus are relevant, including the *dorsomedial nuclei,*[27] which strongly project to the AC, orbitofrontal, and dorsolateral prefrontal cortexes and receive major input from limbic structures in the medial temporal lobe. These thalamic nuclei activate dorsolateral and anteromedial prefrontal structures, as has been shown by functional imaging in patients with selective lesions of these nuclei caused by small thalamic artery infarcts.[28] At the behavioral level patients showed poor ability to abstract, poor insight and judgment, decreased spontaneity and initiative, and impaired executive functions (Kuwert 1991; Sandson 1991; Szelies 1991).

Only bilateral paramedial thalamic lesions[29] lead to coma or impaired alertness or arousal, which is almost never permanent[30] and resolves into an amnesic state[31] or dementia, or severe mental slowing, decreased attention, slow ideation, and apathy. These signs are related to the limbic and prefrontal circuitry and resemble damage to the prefrontal lobes.

The Prefrontal Level. The PFC of the frontal lobe forms the most rostral portion of the building blocks of the PCS. The PCS emanates along the neuraxis from the pontine and midbrain reticular formation, thalamus, hypothalamus, and limbic system, up to the PFC; it is responsible for arousal and alertness, attention, working memory, and executive functions and appears to be the general integration and global control component. It ties all informational threads together; computes dynamic representations of the self, its environment, and possible actions; makes final decisions about whether to act and what action to select;[32] and controls the execution of the chosen action. The PFC, therefore, can be viewed as the keystone component that brings about consciousness in its aware and acting form. To fulfill this task it requires a complex neuronal architecture and connectivity with other major components of the nervous system, including the ILN, NRT, dorsomedial thalamic nuclei, limbic system, and cortical areas. One portion of the PFC, the *dorsolateral* PFC, has to do with *working memory* and allows it to keep a limited number of items in consciousness for immediate use in the performance of ongoing tasks (information is being held online to enable computations to be performed on it) (Fuster 1980).[33] Another important prefrontal portion is the *medioventral cortex* (MV), including the basal forebrain and AC.

The AC forms the *anterior attentional (executive) system* together with parts of the basal ganglia, whereas the *posterior attentional system* includes the *parietal* areas (especially the inferior parietal lobule [IPL] as part of the SCS; see the next section). Owing to its rich reciprocal connectivity with the limbic, motor, and association (especially parietal) cortexes, the AC possesses the informational basis not only for integration[34] of drive and motivational states with sensory and perceptual inputs and motor processes, but also for online monitoring of performance, including selection of targets from among competing inputs and control of alternative responses (forming intentions, decision making, response selection) (Posner and Raichle 1994). Event-related brain potential studies during speeded-response tasks have reported error-related negativity (ERN) with a medial frontal generator in the area of AC, which peaks 100 to 150 milliseconds after an incorrect response is initiated. In this task, AC appears to be associated with monitoring and correcting errors by comparing the representation of an intended correct response with the representation of the actual response (Gehring 1993). The ERN, however, shows up only when the subject identifies the response as incorrect, that is, becomes aware of its incorrectness—an "Aha" experience. An event-related fMRI study confirmed AC's special role in control under conditions of increased response competition (Carter et al. 1998). Some recent data from our own laboratory demonstrate that error-related negativity is lost in patients with circumscribed lesions of the AC. They do not become aware of whether or not their responses are correct.[35]

Taken together, the prefrontal lobes appear to play the most important role in consciousness and to be associated with general online control, including monitoring of internal and external events, prioritization, and planning and correcting (adapting) of behavior, especially if it is novel and complex. Their level of activity increases with the complexity and novelty of the task and decreases with practice (automation).

The Secondary Consciousness System

The SCS is dedicated to specific processing and storage of sensory and perceptual information and provides the PCS with the sensory and perceptual information necessary for computation of aware representations of self and the environment (qualia information). It operates in an automatic and unaware mode.

Sensory input to the cortex is controlled by the posterior attention system, which directs attention to sensory stimuli in various locations in space and includes the parietal cortex, pulvinar, superior colliculus, and NRT, with its gating and perceptual linking mechanisms. Sensory information, perceptions, and concepts are then processed and represented in the modality-specific and heteromodal cortexes of the temporal, parietal, occipital, and frontal lobes and form implicit unaware (neocortical) memories. Information from various sensory cortexes converges in the IPL[36] for polymodal integration and to be passed on to the limbic system for emotional or affective coloring and to the prefrontal-hippocampal system for the establishment of aware episodes and aware memories. Although sensory representations, perceptions, and concepts are modified by experience, they do not give rise to subjective aware experience per se. Rather, *awareness is provided by the integration of perceptual, conceptual, and affective processes and their embedding in the ongoing context of other personality-relevant events in place and time by prefrontal-hippocampal linking.*

The hippocampus has reciprocal connections with both prefrontal and neocortical association areas and receives the outputs of multiple cortical processing systems that can be linked in the process of establishing and retrieving memories. During encoding, awareness is conferred on experience and "woven into the memory trace." Whenever information is within the focus of attention it is consciously apprehended and automatically encoded into memory. The hippocampal-prefrontal memory system appears to be specifically designed to acquire episodes in a consciously accessible format (explicit memory).[37] The hippocampus provides the sense of awareness of memories and allows for re-experiencing, while the PFC plays a critical role in monitoring and verification of memories, for example, in determining whether they are in the appropriate biographical or historical context.[38] Thus the hippocampus and PFC cooperate to produce veridical recollection of aware memories. Hippocampal lesions result in a deficit in aware (experienced) memory (amnesia), whereas damage to the PFC leads to confusion or confabulation. Yet unaware, implicit memories remain preserved in both cases. Correspondingly, sensory information, perceptions, and concepts cannot become aware when they fail to reach the hippocampal-prefrontal structures owing to IPL damage or disruption of IPL-hippocampal-prefrontal connections. Syndromes of unawareness are the result.

Thus the prefrontal lobes appear to be the major integrative component of the PCS and SCS, necessary for the continuous generation of coherent and aware representations of self and the environment.

Clinical Syndromes of Impaired Consciousness and Its Fractionation

In this section major clinical syndromes of impaired consciousness are presented along with their clinical features, underlying neuroanatomy, and implications for the fractionation of the PCS and SCS. They include coma, vegetative state, akinetic mutism, minimally conscious state, and alterations of self-awareness. These labels represent typical and clinically prominent constellations of features due to common underlying pathology and lesion sites.[39]

Disorders of the Primary Consciousness System

Coma. "When somebody collapses, shows only infrequent episodes of respiratory movements, does not move spontaneously, does not respond with speech or other movements to either verbal input or touch or pain and lies with closed eyes, almost every lay person would call him comatose." These features indeed define what medically is termed *coma:* the loss of consciousness, a state of *unarousable unresponsiveness.* There is *no evidence of awareness of the self or the environment* (Plum and Posner 1980, 134). Additional features are revealed by the medical examination (behavioral information) and the CT scan (structural and etiological information).

The patient's pupils are 4 mm and unreactive to light, and all other cranial nerve reflexes are absent, including corneal reflexes. On painful stimulation the patient shows decerebrate posturing. His eyes are continuously closed; no sleep-wake cycles can be observed. On the initial CT scan an intracerebral mass hemorrhage is found, extending into the third ventricle and rostral mid-

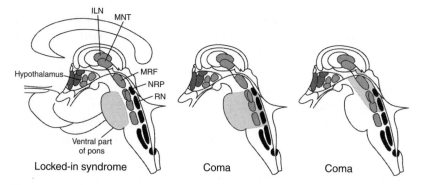

Figure 12.4 The reticular formation and lesion sites relevant for disorders of consciousness. ILN, intralaminar nuclei of the thalamus; MNT, median nuclei of the thalamus; MRF, mesencephalic reticular formation; NRP, nuclearis reticularis pontis; RN, Raphe nuclei. (Modified from Zschocke 1995, 400.)

brain. Except for basic reflexes (which can be spared in shallower forms of coma) such as the cranial nerve reflexes, including pupillary and corneal responses, all brain functions as observable on the behavioral level are severely affected or nil. These include alertness, awareness of self and the environment, affect, emotion, motivation, perception, cognition with attention, memory, language, and goal-oriented movements and behavior.[40]

The case demonstrates the complete loss of all functions in coma, apart from some reflexes, indicating failure of the brain at both its cortical and subcortical levels. Coma was caused in this patient by a massive supratentorial hemorrhage extending to and affecting the midbrain. Coma can also result from widespread bilateral cortical or white matter damage, from intrinsic focal brain stem lesions that disrupt the (rostral part of the) ascending reticular activating system (ARAS),[41] and from extrinsic mass lesions that compress the brain stem. The common end point in inducing coma appears to be impairment of the arousal mechanism by damage to the ARAS in the rostral dorsomedial pons, midbrain, and paramedian thalamus (figure 12.4).[42]

In the case of lesions, which involve only the ARAS and leave other portions of the brain unaffected, an unarousable state results in which all other brain functions are intact but cannot be activated. Given this situation, the medical definition of coma as a *state of unarousable unresponsiveness without any evidence of awareness of the self or the environment* appears to be highly problematic. Although the first part of the definition ("unarousable unresponsiveness") can be tested in clinical examination, "awareness of the self or the environment" remains obscure, as it is clinically not accessible in a person who is unable to communicate his subjective experience. Clinical coma diagnosis, therefore, is always based on the *level* of arousable responsiveness under a stimulus-response paradigm (e.g., the Glasgow Coma Scale). If a patient does not show any responses to painful stimuli, he is classified as deeply comatose.[43] The "awareness of the self or the environment" part of the definition, however, is a subjective interpretation of how the examiner experiences the comatose patient.

Vegetative State. Provided the patient does not die, coma rarely persists for longer than four to six weeks, after which time there is either recovery of consciousness or transition into a vegetative state (VS) (Multi-Society Task Force on PVS 1994; American Congress of Rehabilitation Medicine 1995), as was the case with the patient presented here:

Five weeks after onset the patient opened his eyes but did not withdraw limbs to painful stimuli. He did not follow commands, show any purposeful behavior, or follow visual tracking with his eyes. Six months postonset, the sleep-wake cycle normalized and blink reflexes to auditory, visual, and tactile stimuli could be elicited, but he did not show habituation. No orienting eye movements to loud noise or smooth pursuit could be observed.

In VS patients can be awake (as indicated by open eyes), have normal sleep-wake cycles and can be aroused from sleep. Movements only occur as simple, automatic, stereotypic reactions to stimuli, patients cannot interact with the environment or perform any action that requires planning and appears to be meaningful and goal oriented (figure 12.5). No evidence of cognition or awareness of self or the environment can be observed. There is also no evidence of verbal or gestural communication (Multi-Society Task Force on PVS 1994). Autonomous (i.e., vegetative) functions (cardiovascular, respiratory, metabolic control) and cranial nerve and spinal reflexes are intact in most VS patients.

Functionally, the pathophysiological mechanism in VS is a dissociation of arousal and alertness from the other building blocks of consciousness. When coma resolves and arousal recovers together with

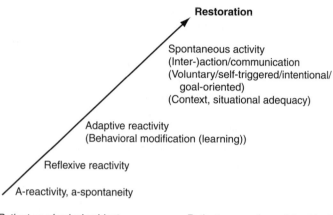

Restoration

Spontaneous activity
(Inter-)action/communication
(Voluntary/self-triggered/intentional/
    goal-oriented)
(Context, situational adequacy)

Adaptive reactivity
(Behavioral modification (learning))

Reflexive reactivity

A-reactivity, a-spontaneity

Patient as physical subject   *Patient as psychosocial subject*

Figure 12.5 Functional aspects of coma remission with respect to the
(re-)activity-action dimension, from areactivity to spontaneous action. In coma
the patient represents an inactive physical subject who develops into an
independent psychosocial subject.

autonomic control, recovery does not progress to the other anatomical
components of the PCS and/or the SCS, and therefore awareness cannot be established (arousal/awareness dissociation or "empty" consciousness). The relevant anatomical components may have been
destroyed, or it may be that they cannot be activated by the arousal
system. Actually several lesion topographies can occur: (1) *neocortical
damage*[44] with involvement of hippocampus, leaving the brainstem
relatively intact; (2) *subcortical white matter damage*[45] with cortical
isolation, and (3) *thalamic damage,*[46] especially of the ILN and RTN,
with loss of cortical activation, sensory gating, and linking.

In (1) both pathologies severely affect awareness owing to missing
content from actual sensory input and processing and from memory.
All acquired information (qualia) is completely lost through neocortical memory impairment and impairment of the encoding and recall
mechanisms. No representation of the self and the environment can be
built up and updated in biographical continuity. Damage to the NRT
gating mechanism in (3) impedes the sensory influx into cortical areas
for further processing and finally for feeding into awareness.[47] In (2),
cortical sensory deprivation also results as in (3), but it is likely to be
permanent owing to the structural damage to the cortical afferent
input pathways in the white matter.

In the majority of coma cases developing into VS, only arousal and the vegetative or autonomous system recover, owing to their close anatomical proximity. However, recovery of some additional building blocks may be seen in a few VS cases. In the course of treating large numbers of VS patients in our specialized unit we have encountered such infrequent variants of the VS syndrome, which include some complex behavior. We observed two VS patients with reproducible emotional responses, such as smiling and laughing to distinct input. In one of them, a twenty-four-year-old male, smiling and laughing occurred reliably whenever his stepfather told him jokes. This behavior is indicative of some linguistic and emotional processing and can be taken as evidence of an at least rudimentary *emotional consciousness*. It may be derived from intact limbic structures acting in combination with brainstem-mediated facial-emotional expression, while the thalamocortical structures are damaged and inoperative, leaving the sensory cortex without input. Despite his remarkable capability, the patient did not recover from VS.

In another VS case complex motor output could be observed. The patient produced motor speech out of any communicative context and in uninterruptible streams of grumbles and intermixed random sequences of phonemes.[48] In addition, his whole body was hyperkinetic-hyperactive, thrashing about like a baby's without performing any useful action. When brought into and held in an upright stance, he performed bipedal movements resembling locomotion. He did not show eye tracking movements and otherwise fulfilled the VS criteria.

It is evident from these cases that when arousal has been re-established other functional systems—such as the emotional, speech motor,[49] and/or head-trunk-extremity motor systems—can become operational if their underlying neuronal structures are preserved and reconnected with the arousal system.[50] Such operational components may demonstrate isolated particular forms of consciousness (e.g., emotional consciousness).

The preceding cases were observable only because the intact emotional processing was connected with the motor system and could be expressed as overt behavior. If, however, motor expression is not available for some preserved (e.g., cognitive) capabilities, they would go unnoticed in clinical examination.

To test this possibility we employed an event-related potential technique to detect a phenomenon known as N 400. N 400 is an electrocortical negative potential observed 400 milliseconds after a semantically anomalous sentence (e.g., "The coffee is too hot to fly") has been heard by a healthy subject. Approximately 10 percent of our patients with a definitive clinical diagnosis of VS exhibit N 400 potentials. This implies that semantic processing and language comprehension take place and therefore that some underlying anatomical structure must be preserved. This may be the case when most of the cortex remains intact, as in the case of cortical sensory deprivation in (2), but deprivation is not complete or when some cortical islands and their input are preserved.[51] The question arises whether preserved semantic processing is an index of consciousness in these VS patients. It is most likely that semantic processing—in the sense of meaning extraction and attribution to perceived information (transformation of sensations and perceptions into experienced phenomena)—is an essential process in becoming and being aware of something. However, semantic processing per se is automatic in normal comprehension.[52] Only the message—the sense—of what has been said reaches the aware consciousness. As N 400 requires not only local semantic processing but also global integration over the whole sentence to extract the message, awareness must covertly exist in at least some rudimentary way in VS when N 400 is preserved.

From these observations it must be concluded that some VS patients exist who are aware in the sense discussed and who exhibit some form of *linguistic consciousness,* but who appear clinically and behaviorally to be in VS. They constitute a variant of VS in which a linguistic component is intact, in addition to the arousal and vegetative or autonomous components; however, it is dissociated from its motor output system.[53]

Akinetic Mutism. Akinetic mutism (AM) is a syndrome denoting *motionless consciousness:* a condition in which patients are awake, appear attentive in that their eyes follow moving stimuli, but cannot otherwise move or speak.

A comprehensive description was given in the first case report. Cairns et al. (1941; Cairns 1952) described the case of a fourteen-year-old girl with an epidermoid cyst of the third ventricle who presented the following symptom constellation:

In the fully developed state she lay alert except that her eyes followed the movement of objects or could be diverted by sound. A painful stimulus would produce reflex withdrawal of a limb but without tears, noise or other sign of pain or displeasure. . . . There were bilateral signs of pyramidal tract involvement, and she was totally incontinent. As one approached her bedside her steady gaze seemed to promise speech but no sound was produced. She showed . . . loss of feeling tone, loss of emotional expression, of spontaneous and most other voluntary movements. She was incapable of originating movements of any kind, with the notable exception that ocular fixation and movement occurred in response to the movement of external objects and to sounds. (1952, 135–36)

Sleep and wake cycles were present, with excessive sleep phases, in which in-principle arousability from sleep was preserved when the patient was stimulated. After the cyst was drained her behavior normalized, but she demonstrated an anterograde amnesia for the interval of AM. Therefore, AM must be associated with an impairment in learning.

When patients recover from AM they demonstrate fluctuating behavior that includes the ability to follow commands or initiate movements and to comprehend speech. Finally, motor speech appears and mutism dissolves (Gugliotta et al. 1989).

AM lesions indicate a breakdown in the PCS along the neuraxis from the paramedian midbrain-diencephalon to the medial forebrain bundle to the PFC; they result in an impairment at the integration point of the PCS, the SCS, and the action-motor system, which varies somewhat with the anatomical structures involved.[54] As a consequence, the prefrontal lobes cannot be activated to intend and initiate motor output, and akinesia results at the behavioral level in the face of preserved arousal (dissociation of consciousness from the motor system).[55] Although some AM patients look highly attentive, their awareness of self and the environment appears to be reduced or nil when the thalamic gating mechanisms in the RTN are affected and sensory input is not provided to the sensory cortex.[56] Perceptions cannot be constructed and are not passed on to the PFC as the informational basis for qualia formation of self and environment (attentive consciousness, dissociation of consciousness from the perceptual system). Emotion and affect are blunted in all AM patients owing to prefrontal isolation from the limbic-temporal-hypothalamic system or lesions in these structures (dissociation of consciousness from the emotional-affective system).[57]

Whereas AM represents an impairment at the integration point of the PCS, the SCS, and the action or motor system, which varies somewhat with the anatomical structures involved, a lower-level interruption in the motor system occurs in locked-in syndrome (LIS), which mimics AM. Damage to the motor output pathways in the anterior pontine brain stem produces akinesia (except for vertical eye movements) and mutism, yet leaves the structures underlying consciousness fully unaffected. It is, however, unclear how the loss of movement and the experience of an immobile body influence self-awareness.

Minimally Conscious State. When patients demonstrate fluctuating but reproducible behavioral evidence of self-awareness or awareness of the environment but are unable to follow instructions or communicate reliably, the term *minimally conscious state* (MCS)[58] is applied as a diagnostic label (Giacino and Kalmar in press; Giacino et al. in press). In MCS, cognitively mediated responses occur inconsistently and often depend on external stimuli. In this context, consciousness is determined clinically, with various degrees of diagnostic certainty, along a range of features ranging from visual fixation, visual tracking, and inconsistent emotional or motor behaviors that can be elicited reliably and distinctively by specific stimuli (lower certainty of the presence of consciousness) to inconsistent command-following, object manipulation, or communication through intelligible verbalization or gestural or verbal yes/no responses (higher certainty of the presence of consciousness). Establishment of full consciousness with linkage to the motor output or action system is documented by consistency of behavior in command-following, functional object use, and reliable use of communicative systems.[59]

The underlying neuropathology of MCS is most likely related to diffuse, bilateral, subcortical and hemispheric damage that is less complete compared to that underlying VS and that varies from case to case. Some cortical islands and sensory afferent pathways may have been preserved and may become operative again when NRT gating and binding mechanisms in the thalamus recover and allow sensory information to reach the cortex for computation of self- and environment representations. This recovery would signal a restoration of the SCS.

Disorders of the Secondary Consciousness System

In this section disorders of the SCS are presented, including some syndromes that fall on the borderline between the SCS and the PCS. They are related to impairment of the PFC, which appears to be the keystone for integration of both systems.

When patients suffer from lesions to the SCS they remain awake and interactive with their environment but exhibit a variety of disorders that affect awareness of self or environment. Perception, or some aspects of it, may basically remain intact to allow for implicit, unaware processing (dissociation between perception and awareness). Such disorders include hemineglect, blind sight, cortical blindness, agnosia, anosognosia (unawareness of deficits), and asomatagnosia (personal unawareness). Confabulations can occur in cases of disintegration of the PCS and SCS at the prefrontal level.

In *hemineglect* patients are unaware of one body side or one hemispace in the environment. They do not respond to stimuli from intra- or extrapersonal[60] space contralateral to the lesion site, despite preserved primary sensory processing in the visual, auditory, and tactile domains. Multimodal hemineglect results from unilateral paramedian thalamic lesions or from damage to the IPL. In both cases, sensory integration is impaired and stimuli cannot become aware as they are not passed on to PFC/hippocampal structures. The latter is also the case when the IPL is damaged. Unimodal hemineglect occurs when information from a sensory (e.g., visual) area does not reach the IPL.

The representation of the body (*body image*) may be distorted when primary sensory areas are disrupted from parietal regions. Dysfunction of primary and association areas results in hemisensory loss and various other syndromes in which stimuli or aspects of them can be processed implicitly but remain unconscious. Among these are blind sight, "blind" touch, various forms of *agnosia,* or a failure in perception with preserved sensation, as in *prosopagnosia,*[61] *simultanagnosia,*[62] or *Balint syndrome.*[63] In addition to hemineglect, *asomatognosia* may occur after right supramarginal and posterior corona radiata lesions in which a patient denies ownership of, say, his left arm owing to a loss of its relatedness to the body. In *anosognosia* a patient denies or is unaware of illness, for example, that her arm and leg are paralyzed.[64] Both disorders may occur separately, as when a patient admits that he had a stroke with left hemiparalysis yet believes that his left arm and leg do not belong to him.[65]

These disorders often include damage to the prefrontal lobes, including basal forebrain and anterior cingulate (AC) areas, as do more complex alterations of self with withdrawal of personal relatedness, as in Capgras syndrome, which represents a delusional belief that a person—usually someone well known to the patient—has been replaced by a double or an impostor. An insertion of personal relatedness occurs in cases of Frégoli syndrome: the patient exhibits a persistent belief that someone who is actually unfamiliar to her has taken the psychological identity of someone she knows.

Disintegration of self- and environmental awareness is found in prefrontal patients who live simultaneously in two worlds, for example, at home and in the hospital (double-life syndrome; Schoenle unpublished cases). Some patients shift from one environment to the other and live sequentially in the two. Others experience both environments simultaneously; often the room they are in is felt to be one environment and the outside is perceived as the other, for example, the marketplace in their hometown. Some patients are aware of the fusion but are unable to correct it. The double-life syndrome may result from degraded control and coordination of two informational lines, one that consists of experienced memories conferred by the hippocampus and another that comprises the experienced present. Both streams, of past and present biographical information, are experienced as aware and coexisting simultaneously or sequentially. In other cases aware, experienced memories flow in from the hippocampus completely uncontrolled, to dominate and overlay the experienced present. Such a patient may believe he is at work and that the people around him are colleagues. In extreme cases, these patients can be approached meaningfully (e.g., for psychometric testing) only from within their delusional systems.

Another type of disintegration of self and environment can occur in confusional states.[66] Patients do not know where they are, what time it is (date, time of day, season, day of the week), or what situation they are in; they may also not know their marital status, may not recognize their spouse or children, or know what job they hold. Some even lose the appreciation of self and qualities of self-experience to such a degree that their sexual identity is unclear to them and they are uncertain whether they are male or female. Self-awareness can be distorted even to the point that identification of self is no longer possible (disautia). We recently observed a patient of ours who was on a national television program. When he watched the program he did not recognize himself. When asked who this person was, he thought it was some other person from work.

Confabulations are nonveridical memories sometimes intermixed with veridical memories, which are taken out of their biographical context and miscombined; they are sometimes bizarre and often persistent—despite their self-evident implausibility—and are comparable to convictions seen in patients with delusions. They may be due to impaired frontal lobe monitoring and verification of memories that are provided by the intact hippocampus as aware and experienced memories.

When patients cannot integrate perceptions of the environment into aware, coherent representations of the environment, they are unable to perform spontaneous, self-triggered actions in that environment. In one such patient we could demonstrate that she was completely unable to act spontaneously in real-life environments (for example, on her ward) but was perfectly able to play a game like ludo. (In this game the construction of the mental representation of the environment is rule-based, as is the action from moment to moment during the game, even if the players are new and the way in which they act is not known from prior experience.)

This observation demonstrates a crucial role of AC structures in online integration of environmental and self-representations for the control and implementation of "free" actions carried out by the self on or in the environment, which is also "free" (in the sense of not being an automatic behavior in a well-known environment).[67] The patient had suffered a ruptured anterior communicating artery aneurysm with a bilateral focal lesion in the AC.[68] Often the extent and constellation of behavioral impairments vary from patient to patient after ruptured-aneurysm bleeding in the anterior communicating artery, depending on which of the prefrontal and basal forebrain structures are affected.

Conclusion

From the preceding survey of the consequences of brain damage for consciousness, it is obvious that consciousness is not a unitary phenomenon. Analysis of disorders of consciousness and their underlying neuroanatomical substrates allows the gross division of the structural and functional components of human consciousness into two basic systems. The PCS represents the general-purpose processing system, which provides arousal and activation, attention, working memory, and executive functions to finalize action. It extends from the caudal brain stem and MRF, through paramedian mesodiencephalic structures (NRT and ILN) to the PFC. The SCS is based on mainly thalamosensory-cortical structures and is dedicated to specific sensory-perceptual processing and storage. It provides information about the body and the outside world to form the qualia of self and the environment. Depending on the lesion site and extent of damage, further fragmentation may occur within both systems owing to brain injury, resulting in distinct alterations of consciousness (e.g., in hemineglect) or in narrowing of consciousness into a vegetative, emotional, linguistic, or attentive variant.

Disorders of the PCS range from coma, VS, AM, and hypersomnia to low-consciousness states. They are variants of PCS disorders, due to smaller, circumscribed lesions that fractionate the primary system.[69] They fulfill the *third prediction,* that a breakdown of the PCS leads to unavailability of the SCS. Focal coma results when PCS is affected at its lower level in the anterior and posterior pontine and mesencephalic brain stem. As a consequence, thalamic structures, including ILN and NRT, are not activated, and therefore information is not gated to the cortex, and cortical activation and cortical binding do not occur. Thus even if the SCS is not damaged structurally and the qualia information exists and is potentially available, it cannot be read out from its cortical representations as long as these brain stem areas are affected and the comatose state prevails.

When these brain stem areas recover, *arousal* returns in VS patients, who remain, however, unable to activate cortical information of the SCS system through thalamocortical circuits due to ILN and RTN lesions. These patients may have completely intact cortexes (e.g., in the left hemisphere) with potentially preserved linguistic knowledge, which, however, cannot be read out and transformed into aware information, as indicated by a preserved N 400 response. In some AM cases with prefrontal impairment, SCS may also be intact but disconnected from the PFC, leaving the patients in full arousal and attentive, yet perceptually uniformed. Intactness of the SCS can easily be demonstrated clinically in hypersomnia due to lesions restricted to the rostral mesencephalon, as patients can be aroused to awareness and full responsiveness.

The *second prediction* of an impairment of the SCS is supported by VS patients with devastating cortical brain damage in which the cortex is completely or almost completely destroyed in both hemispheres. This state was originally (before *vegetative* was introduced) termed *apallic*

(Gerstenbrand 1967), meaning "without pallium," that is, without any remaining neurons in the cortex. As a consequence patients are devoid of any cortical information and cannot construct conscious qualia despite normal arousal. A similar situation is found in Alzheimer's disease, when all cortical neurons have been destroyed except those in the primary motor and sensory areas. Alzheimer's patients, therefore, cannot form aware representations of self and the environment, but have—at least in the disease's earlier stages—no motor or sensory restrictions: they can walk around, speak, see, hear, and feel.

A fragmentation of the SCS can occur with focal lesions in which primary or associative cortical areas are damaged or connections to prefrontal or hippocampal structures are interrupted. Unaware, implicit processing may still be possible if some aspects of sensory or perceptual representations are preserved, as is the case in blind sight, hemisensory loss, hemifield defects, agnosias, or hemineglect. In these cases, patients are fully awake and interact with the environment but exhibit some partial unawareness of self or the environment.

The *first prediction* states that the PCS and SCS are intact but are not integrated adequately. This appears to be an extreme prediction in that it calls for completely preserved primary and secondary systems, a combination that may be very rare in reality. Yet it postulates an integrative function that may be valuable in explaining cases in which disintegration plays a dominant role, even if other aspects of consciousness may also be affected to some extent.

Disintegration of both systems may be limited to some aspects of consciousness, as in patients with multimodal hemineglect with unawareness of half a body or half of the environment due to damage to the IPL and disconnection from prefrontal and hippocampal structures.

Disintegration, however, may be far more profound when the prefrontal lobes are damaged.[70] Patients are awake; they can speak, listen, and understand and execute instructions. They cannot, however, tie together incoming information about the present and information about the experienced past to construct a continuous and coherent representation of self and the environment. Manifestations range from confusion and confabulations to delusions, reduplication of self, double-life syndrome, disintegration of self (disautia), alien hand syndrome, and Capgras and Frégoli syndromes.

The prefrontal cortex appears to be the cornerstone of consciousness. Even small bilateral focal damage to the AC in the prefrontal lobe has devastating effects on the human ability to operate as an open, *self-aware, self-learning,* and *self-organizing* system that can cope with novel environments and situations, as was the case with the patient described earlier. She could not abstract a representational model of the world from the outer, concrete, changing environment; compute a representation of the self embedded within the world model; and thereby constitute awareness of her self and her environment. As a consequence, she was unable to compute models for intervention, to project these onto her environment, and to select action models for implementation. Therefore she could not perform spontaneous environmentally adequate actions and would not have been able to survive had these deficits not been compensated for in a sheltered environment. Consciousness, in this ultimate sense, is the action control system for our survival.

Taken together, data from patients with brain damage suggest a multimodular framework of consciousness. It encapsulates multiple components along the caudal-rostral neuraxis and unfolds at various levels for integration of additional functionality, ranging from vegetative (brainstem component) to emotional-affective (limbic component), inner state-drive (hypothalamic component), and cognitive (cortical component) functions. The final synthesis takes place at the prefrontal end of the neuraxis, where all relevant information is integrated and formed into aware representations of the self and the environment[71] to allow the self to act adequately on or in its environment.[72] In brain damage the components can be distinctively affected, preserved, or dissociated from each other and can surface as partial manifestations (such as vegetative, emotional, or linguistic consciousness) of the normally integrated multiple-consciousness system.

The task now facing those who adopt a neurological approach to consciousness will be to further describe the phenomena[73] of altered states of consciousness at the structural-neuroanatomical, neurochemical, neuropharmacological, and functional levels, employing behavioral observation, structural and functional MRI, MRI spectroscopy, and PET as well as neurophysiological investigations at various levels of complexity. It will be of special interest to study the interactive dynamics of the multiple functional modules, their disintegration and reintegration in bringing consciousness into existence from moment to moment, and the role of the PFC in orchestrating the multiple consciousness systems. The challenge will be to give a coherent account of the multiple alterations of consciousness in patients with brain damage, and, even more, to explain why consciousness recovers to the extent it does in one patient but not in others. The ultimate goal is to arrive at more effective intervention strategies to achieve improved regeneration of consciousness.[74]

NOTES

An extended version of this chapter, covering aspects that could not be detailed here, appears in Schoenle (in press).

1. Many of the details (and especially the neuroanatomical and neurological nomenclature) are not covered, as they are readily available in standard medical textbooks (e.g., Nieuwenhuys et al. 1991).

2. Pathological phenomena arise when functional systems (e.g., the liver, kidney, brain) are altered by some etiology (e.g., virus infection). A medical (i.e., physiological) theory that describes and explains the normal functioning of such a system can be tested with respect to its validity and power when it is confronted with pathological phenomena. If a theory of the liver or another organ is valid, it should be able to explain how pathological phenomena arise and are caused by a given etiology (e.g., a virus) and how interventions can be deduced from the theory. The explanatory power of the physiological theory of normal functioning is proportional to its capability to integrate pathological phenomena, that is, to become a theory of pathophysiology (a valid but weak version of the theory), and its usefulness in providing therapeutic means (a strong version of the theory).

3. We had the opportunity to open the first center and model unit for ICR in the state of Baden-Württemberg ten years ago. It includes an affiliated unit for long-term care of patients who show no further improvement after six months of rehabilitation. Funding was provided by the Social Ministry of Baden-Württemberg.

4. The portion of the NR approach discussed here is focused on functional restitution and represents a somewhat reductionist third-person perspective, whereas NR as a whole is humanistic and takes a *we* perspective. It attempts to meet the patient within the subjective experience of his deficits and reduced degrees of freedom, in his struggle with coping and self-acceptance. It accompanies him in developing a new life perspective and in achieving as much life satisfaction as possible given his remaining capacities.

5. The ultimate goal from this perspective, of course, is to ensure phylogenetic survival.

6. In the sense that it is presented and present to the brain. Representations allow the re-instantiation (re-representation) of formerly present worlds and the computation of representations of possible future worlds.

7. Part of the abstraction process is the ability to use symbols, most importantly language, as a means for construction of representations and virtualization. The loss of this ability to use symbols is called asymbolia.

8. The external world is meant to include not only the physical but also the social environment.

9. The most important code appears to be language. It allows us not only to build and process inner representations (inner speech) but also to transfer these representational worlds to other persons, even those in future generations (language as a noetic code for the transmission of cultural information comparable to DNA as the biological code for genetic information).

10. Updating of sensory input may occur over 300- to 500-millisecond periods—the time quanta the brain requires to transform sensations into perceptions (Libet 1978).

11. This is why hypoglycemia decreases concentration and learning performance and why drugs that block intracellular protein synthesis impede or block attention and learning. These neuronal systems are also very sensitive to hypoxemia and are thus severely affected during cardiac arrest, drowning, or status asthmaticus.

12. Abstraction and representation allow for virtualization of the external world and subsequent virtual manipulation (modeling) before acting. Another aspect that is pertinent but not detailed here is the loss of the ability to use metaphors and to construct sense and meaning after brain damage. Both abilities are, from an evolutionary point of view, major achievements (development from sensorimotor brain to action brain through conscious brain: representational brain, virtual brain, and metaphorical brain).

13. These include neuroimaging (computed tomography [CT], MRI, PET) and neurophysiological, neurochemical, neuroimmunological, and neurogenetic studies.

14. The prefix *neuro-* in *neurofunctional* is intended to convey that functions of the central nervous system are the object of the analysis.

15. Acute neurology is more structurally oriented. What caused the alteration of the structure underlying the dysfunction or loss of function? What are the mechanisms causing the damage? In general, the structure-function dichotomy is artificial and linguistic in nature. Physiologically the dichotomy does not exist: building up a cellular or subcellular structure and preserving it are per se the prime functional outputs of that very structure (its intrinsic functions). Extrinsic functions (e.g., exciting other cells for the purpose of information transmission) are additional outputs. In the case of alterations these extrinsic functions are the first to be compromised or supported. For example, in the case of hypoxia, (extrinsic) functions are lost before the structure (intrinsic functions) dissolves (functional metabolism versus structural metabolism).

16. The function-analytic approach is a precondition for localization of function in the neuronal substrate—be it through functional imaging or lesion localization. In pathological cases we must first perform a thorough and comprehensive analysis of the functional components in order to identify the compromised component. Then we are ready to look for the underlying locus of the lesion and

attempt a correlation (functional localization precedes brain localization). In functional imaging studies of non-brain-damaged subjects a similar approach is warranted. Owing to lack of space I do not detail here the double dissociation approach for functional lesion localization. The epistemological question, however, is whether the functional decomposition, the levels of observation, and the language of description are the same ones that the brain uses to organize itself. Most likely the hardware code is not identical with our own user code.

17. Although it is suggested that there are two principal consciousness systems (CSS), they are not mutually exclusive. The dichotomy is offered as a framework within which the various issues affecting CSS may be organized. There are, of course, various levels of description of consciousness: the abstract level of functional analysis (abstracting from the physical implementation), the description of the structural architecture of the nervous system related to consciousness (one implementation among many possible ones), and the description of its functional architecture (the dynamic operational mode of the interacting components of the system).

18. Attention and working memory may be thought of as a development and extension of the arousal-alertness component.

19. However, there are very small areas with great functional significance that can have devastating effects when lesioned, including those in the thalamus or the pontine or mesencephalic brain stem.

20. It is conceivable that such modules could be members of several virtual work groups. This scheme, however, would place high demands on coordination or management, a function that could be mediated by binding and performed by thalamic pacemakers. Minor disadvantages would be higher sensitivity to interference (e.g., of drugs) on the polysynaptic infrastructure and longer processing time. Therefore, permanent work groups appear to be formed with optimized connectivity whenever specific tasks must be performed over longer periods of time. These are functional systems (e.g., the respiratory control system). This type of functional organization is called dynamic task-dependent distributed networking.

21. The term *reticular* alludes to the netlike appearance of the neurons under Golgi staining. The RF that relates to consciousness is located mostly in the paramedian tegmentum of the pons and midbrain on its way to the intralaminar thalamic nuclei. It courses through the tegmental regions on both sides of the Sylvian aqueduct and the banks of the third ventricle (Caplan 1996, 357).

22. A loss of LC neurons occurs in Alzheimer's and Parkinson's diseases.

23. *Afferent* refers to the sensory input from the body to the RF and *efferent* to the output from the RF to the motor system.

24. The cranial nerve nuclei mediate sensation and motor activity in the head and the face area, also subserving speaking, chewing, and swallowing.

25. The thalamic midline and intralaminar nuclei are the so-called "nonspecific" cell groups of the thalamus.

26. The dorsal leaf of fibers runs to ILN, the ventral leaf to NRT laterally and ventrally through the subthalamus and hypothalamus.

27. Other nuclei are the ventral-anterior, the ventrolateral, the medial and anterior pulvinar, and the anterior nuclear group, all strongly interconnected with the frontal lobes.

28. Symptoms of an *anterolateral thalamic (polar) infarct* include apathy; slowness; decreased spontaneity; abulia; lack of emotion and emotional responses; impaired insight; loss of concern, motivation, and memory; and difficulty in learning; they resemble the symptoms of frontal lobe syndrome. Indeed the dorsomedial nucleus and other thalamic nuclei—including VA, VL, the medial and anterior pulvinar, and the anterior nuclear group—have strong reciprocal connections with the frontal lobe. The dorsomedial nucleus strongly projects to the anterior cingulate, orbitofrontal, and dorsolateral frontal cortexes and receives its major input from limbic structures in the medial temporal lobe. A patient with a left polar artery infarct showed decreased single-photon emission computed tomography uptake in the dorsolateral and anteromedial frontal lobe, which functionally was paralleled by decreased spontaneity and initiative, poor insight and judgment, perseveration, poor ability to abstract, poor response inhibition, and impaired executive functions (Sandson 1991; Szelies 1991). PET studies also reveal reduced activity in caudate nucleus and frontal lobe on the side of the medial thalamic infarct in such patients (Kuwert 1991). A patient with a left polar (tuberothalamic) artery infarct and lack of spontaneity and emotional responsiveness was described by Lisovoski et al. (1993, 183): "She showed lack of spontaneity and affective indifference. She did not manifest emotions and showed no emotional concern for her illness. When she was directly stimulated by another person, she was able to act normally. However her activities would stop immediately when the external stimulation disappeared."

29. Including intralaminar, parafascicular, medial dorsal, and ventral parts of the centromedial nuclei (Gentelini et al. 1987).

30. Often lesions extend into the rostral midbrain, with impairment of vertical gaze (Parinaud syndrome).

31. Memory problems are related to bilateral lesions of the mammillothalamic tract or the mediodorsal thalamic nuclei.

32. These processes are dynamic in the sense that they are performed continuously from moment to moment.

33. Prefrontal lesions are associated with disturbances in the ability to spontaneously generate items from a list presented earlier (in free-recall memory tasks), impairment in temporal ordering of recently experienced events, and remembering the source and context of information. Nonmnemonic impairments include the inability to inhibit responses, to ignore irrelevant information, or to inhibit previously correct responses, which could affect search and retrieval strategies necessary in mnemonic processing (Shimamura 1995).

34. The functional architecture that operates over and thereby integrates the components of this architecture (linking) may be based on (among others) the 40-Hz oscillatory processes. These appear to phase-lock and link different distant cortical areas that are dedicated to the processing of distinct aspects of sensory stimuli (Munk 1996). When linking is employed repeatedly, it becomes structuralized (transformed from a functional dynamic organization into a structural

hardware organization) and is thus less susceptible to interference, at the cost of flexibility (a process of automatization and transfer to dedicated processing units, i.e., modules).

35. Lesions in these patients were caused by a hemorrhage from an aneurysm of the anterior communicating artery affecting the AC and the subcallosal and preseptal areas (i.e., the basal forebrain).

36. The IPL has connections to various sensory cortexes, the limbic system, and the PFC.

37. Explicit memory is mediated by limbic mediotemporal lobe structures (hippocampus, amygdala, entorhinal and perirhinal cortex, and parahippcampal cortex) and diencephalic limbic structures (medial thalamic nuclei and mamillary bodies).

38. The PFC is also of importance when effort is required for encoding and retrieval.

39. It must be noted, however, that, aside from these prototype descriptions, a vast variety of other combinations can occur, owing to the high degree of individualization of brain damage. These disorders do not bear special labels, but they are equally important for future discussions of consciousness because they provide further details about the more fine-grained fractionation of its neuronal substrate.

40. In detail, lost functions in coma include alertness/arousal, wakefulness, control of sleep/wake cycles, vigilance, drive, motivation, emotion, volition, affect, perception of interoceptive and exteroceptive stimuli, cognition with attention (focusing, filtering, selection), orientation, memory, language, thinking, deductive reasoning, ideation, abstraction, integration from past and present, verification of perceived and remembered information, modeling, projection, anticipation, evaluation, (moral) insight, planning and selection of actions and behaviors, (spontaneous) purposeful movements (e.g., pointing, localizing defense responses), decision making, control, continuous internal and external supervision and permanent readiness for action, initiation and termination of actions, and adaptive, context-adequate, goal-directed behavior. Reflexes, such as the corneal reflex, are lost only in the severest forms of coma.

41. In the differential diagnosis of coma a vast variety of intra- and extracerebral etiologies must be taken into account, ranging from structural disturbances (including cerebrovascular accidents, infections, epilepsy, tumor, and brain injury) to metabolic and endocrine disorders (including anoxia and electrolyte imbalances) to toxic (including alcohol and other drugs) and pharmacological (including anesthetic and narcotic) influences. In general, whenever the intracranial pressure rises in the skull, the only way in which it can be accommodated is through downward expansion of the brain toward the foramen magnum at the base of the skull, thus damaging the brain stem, with the reticular system, the afferent and efferent pathways, and the cranial nerve cells.

42. Such damage can be caused extrinsically by pressure on the brain stem or intrinsically by infarcts or compression of the branches of the basilar artery supplying the brain stem.

43. If the patient can report his name, the date, and his location when asked, he is considered to be out of coma.

44. Reference to the destruction of the neocortex is made in the term *apallic syndrome*, which implies that the pallium (i.e., the cortex) is missing. Neocortical damage with diffuse laminar cortical necrosis (especially of the larger cell layers 3, 5, and 6) and involvement of the hippocampus (especially areas CA1 and CA4–6) may be due to severe anoxia (e.g., following drowning or heart failure), leaving the brainstem relatively intact.

45. Cortical isolation caused by diffuse axonal injury and subcortical white matter damage develops after traumatic brain injury. Acute trauma leads to shearing, decelerating-accelerating, and rotational forces, which tear white matter axons, with subsequent diffuse axonal degeneration and isolation of the cortex from subcortical structures (Jennett 1991).

46. The thalamus is affected in the paramedian and lateral regions of the central and posterior parts, especially of the "nonspecific" ILN and RTN, including the forniceal system, mammillary bodies, posterior cortical watershed zones, and occipital lobes. Evidence in support of the diencephalic basis of VS comes from the autopsy of the brain of Karen Ann Quinlan, who suffered a cardiopulmonary arrest at the age of twenty-one and survived for almost ten years in VS. Her behavior comprised wake and sleep cycles, spontaneous eye opening, eye opening to auditory stimuli, simple withdrawal-type responses to stimuli, and spinal reflex movements without any goal-directed, purposeful actions (Kinney et al. 1994).

47. When NRT recovers, however, informational content can at least to some extent be provided for awareness, as sensory afferent pathways and cortical processing and storage remain unimpaired in (3). This constellation constitutes a possible mechanism for recovery of self- and environmental awareness in the transition from VS to a minimally conscious state.

48. The speech output was comparable to that observed in phonemic jargon aphasia (PJA). These patients are unaware of their distorted speech which appears to run free and out of their control.

49. Teeth grinding and chewing are more frequently observed orofacial movements.

50. In the case of nonmotor (e.g., cognitive) functions, they must be reconnected not only to the arousal system, but also to some parts of the motor system in order to become manifest at the level of overt behavior.

51. As these patients do not show much in the way of overt behavior, all available technologies must be applied to tap their preserved inner functional capabilities (covert behaviors). This is of special importance in the ethical and legal contexts. In some Western countries these patients may legally have food and water withdrawn and thereby be starved to death, despite the fact that the decision is based only on clinical criteria. VS therefore poses a major ethical problem, because of both the uncertainty in clinical diagnosis and the diagnostic underspecification of functional abilities.

52. Semantic processing (e.g., the computation of semantic relationships within a sentence) is most likely performed automatically.

53. In these cases auditory pathways are intact (for presentation of the language material), but the other sensory afferents may be interrupted, as documented by somatosensory evoked potentials.

54. AM usually occurs with lesions in dorsomedial thalamus, posterior thalamic-rostral mesencephalic lesions, white matter lesions of the frontal lobes, or destruction of the medial or orbitofrontal surfaces of the frontal lobes, especially when the ACC is involved (Fesenmeier et al. 1990). In some cases AM can be reversed by dopaminergic agents (e.g., bromocriptine), pointing to reduced dopaminergic activity (e.g., in the median forebrain bundle or rostral forebrain) as a cause (Milhaud et al. 1993). In principle, two pathophysiological mechanisms may account for AM: impairment of reticular cortical or limbic cortical functions, by and large sparing corticospinal pathways. Small lesions in the paramedian ARAS of the posterior thalamus and adjacent midbrain lead to diminished or zero thalamic (ILN) output, which is responsible for impaired or zero activation of frontal (especially prefrontal) areas and subsequently results in akinesia. Frontal lobe lesions may also directly cause impairment of the motor system. The absence of cognitive functions may be explained by impairment of thalamic gating when damage to RTN impedes sensory influx to the cortex. On the other hand, lesions to the basal forebrain (including ACC) and other limbic circuitry may lead to a deficit in processing of experienced memories. Thus interruption of the limbic input to PFC may affect cognition profoundly, just as interruption of the midbrain ARAS has profound effects on arousal and activity. Patients with frontal mutism appear to be more wakeful than those with thalamic or midbrain lesions, and they sometimes even exhibit hyperactivity in the form of episodically agitated behavior due to an instability of the behavioral control system (Anderson 1992). Hyperkinesia may result from dissociation of the motor system from prefrontal control. We have observed several such patients who were hyperactive and even agitated, although they appeared to be unaware of self and environment.

55. In some cases bilateral damage to the motor output system, including globus pallidus and caudate nucleus, may coexist.

56. In these cases the posterior (thalamoparietal) attention system appears to be impaired; the anterior (prefrontal) attention system may be functioning, yet it is disconnected from the motor output system.

57. A dissociation of emotional awareness and emotional motor output is observed in patients with pathological laughter or crying. Patients cry but report that they are not in the corresponding mood. They stop crying when they become involved in some other activity.

58. MCS has recently also been adopted as the abbreviation for the term *minimally responsive state*, which had previously been defined by the American Congress of Rehabilitation Medicine (1995).

59. This operationalization of consciousness in medical practice is a reduction of consciousness to the action system. It implies that the patient is able to move some body part to externalize information about her internal cognitive capacities—that is, conclusions are drawn about covert behavior from overt behavior. There are, however, neurological conditions in which patients cannot move (e.g., VS, LIS, and AM).

60. Personal neglect results in the first case, spatial neglect in the second.

61. In prosopagnosia the patient fails to recognize familiar faces owing to inferior temporal lobe lesions. He may know that the image is a face and perceive

simple forms, but he cannot integrate sensations with a coherent gestalt and link sensation to memory.

62. In simultanagnosia patients are unable to integrate information into a general shape while parts of the object can be perceived, owing to somewhat larger lesions involving multiple prestriate regions.

63. Balint syndrome results from bilateral parietal lesions and is characterized by (1) constriction of visual attention with the inability to perceive more than one object at a time and (2) optic ataxia, the inability to reach accurately for an object.

64. Unawareness of hemiplegia is also called Anton's syndrome.

65. This indicates that no anosognosia is present. The denial of ownership demonstrates asomatognosia.

66. Acute confusional state syndrome arises from a global disturbance in the integrated function of brain structures involved in alertness and arousal and attention (PCS). Confusional states also occur in the case of focal mediotemporal (hippocampal), inferior parietal, or prefrontal lesions in the right hemisphere.

67. In addition to her inability to act, she also was confused and confabulated.

68. These can affect the basal forebrain, AC, anterior hypothalamus, anterior columns of the fornix, septal nuclei, anterior commissure, and genu of the corpus callosum. As mentioned earlier, patients with anterior communicating artery aneurysmal bleeding exhibit a loss of error-related negativity (ERN). ERN, therefore, appears to be related to monitoring and verification of information emerging from memory versus actual environmental information arriving from the sensory periphery of the body. When these monitoring and verification processes are affected, past and present information intermingles and confabulations result.

69. "Pure" cases like the ones just described apply to the foregoing hypotheses, as they are caused by smaller, circumscribed lesions with distinct effects on the PCS, the SCS, or both. The vast majority of patients, however, suffer from extended, multifocal, or diffuse brain damage due to, for example, hypoxia or severe brain injury. Therefore no differential functional effects result, but rather a global breakdown of both the PCS and the SCS. Yet even in these global cases islands of preserved neuronal tissue may resume activity and lead to isolated behaviors like adequate emotional responses to jokes or isolated sound production without any meaning, as seen in some VS patients.

70. Damage to the prefrontal lobes generally results in lack of spontaneity, initiative, and decision making; poor insight and judgment; confabulation; perseveration; poor ability to abstract; poor response inhibition; and impaired executive functions.

71. Whatever information prevails determines the qualia of consciousness (e.g., cognitive, emotional, or vegetative).

72. Although at each level integration can converge into motor output, its quality varies from reflexive to drive-dependent, emotionally or cognitively mediated, or intended action and behavior.

73. It will be of special interest to find out how consciousness, with its representation of the self, is modified by complete deefferentation in LIS or by deafferentation in some polyneuropathies or bilateral capsula interna or thalamic infarcts. It is most likely that self-representations shrink or even become "dis-

embodied" to the extent that body information is not available. This contrasts with cases of Alzheimer's disease, in which representations—per se—of the environment, the body, or both disappear in consciousness due to decortication. 74. As consciousness is not a unitary phenomenon, the question of whether patients who do not regain consciousness should be denied tube feeding and medical treatment is not only naïve from a neuroscientific viewpoint but also unethical, given the uncertain status of our current clinical investigations of individual patients and the multiplicity and complexity of the phenomenon. Which fragments of consciousness are more valuable? How many fragments must be intact in order for a patient to be allowed to live?

REFERENCES

American Congress of Rehabilitation Medicine. 1995. "Recommendations for Use of Uniform Nomenclature Pertinent to Patients with Severe Alterations in Consciousness." *Archives of Physical Medicine and Rehabilitation* 76: 205–9.

Anderson, B. 1992. "Relief of Akinetic Mutism from Obstructive Hydrocephalus Using Bromocriptine and Ephedrine." *Journal of Neurosurgery* 76: 152–55.

Cairns, H. 1952. "Disturbances of Consciousness with Lesions of the Brain-Stem and Diencephalon." *Brain* 75: 109–46.

Cairns, H., R. C. Oldfield, J. B. Pennybacker, and D. Whitteridge. 1941. "Akinetic Mutism with an Epidermoid Cyst of the Third Ventricle." *Brain* 64: 273–90.

Caplan, L. R. 1980. "Top of the Basilar Syndrome." *Neurology* 30: 72–79.

———. 1996. *Posterior Circulation Disease*. Oxford: Blackwell.

Carter, C. S., T. S. Braver, D. M. Barch, M. M. Botvinick, D. Noll, and J. D. Cohen. 1998. "Anterior Cingulate Cortex, Error Detection, and the Online Monitoring of Performance." *Science* 280: 747–49.

Fesenmeier J. T., R. Kuzniecky, and J. H. Garcia. 1990. "Akinetic Mutism Caused by Bilateral Anterior Cerebral Tuberculous Obliterative Arteritis." *Neurology* 40: 1005–6.

Fuster, J. M. 1980. *The Prefrontal Cortex*. New York: Raven Press.

Gehring, W. I. 1993. "A Neural System for Error Detection and Compensation." *Psychological Sciences* 4: 385–90.

Gentelini, M., E. DeRenzi, and G. Crisi. 1987. "Bilateral Paramedian Thalamus Artery Infarcts: Report of Eight Cases." *Journal of Neurology, Neurosurgery and Psychiatry* 50: 900–904.

Gerstenbrand, F. 1967. *Das traumatische apallische Syndrom*. Berlin: Springer-Verlag.

Giacino, J. T., and K. Kalmar. "The Vegetative and Minimally Conscious States: Comparison of Clinical Features and Functional Outcome during the First Year Post-Injury." *Journal of Head Trauma Rehabilitation* (in press).

Giacino, J. T., N. D. Zasler, D. I. Katz, I. P. Kelly, J. H. Rosenberg, and C. M. Filley. "Practice Guidelines for Assessment and Management of the Vegetative and Minimally Conscious States: Proceedings of the Aspen Neurobehavioral Conference." *Journal of Head Trauma Rehabilitation* (in press).

Gugliotta, M. A., R. Silvestri, P. De Domenico, et al. 1989. "Spontaneous Bilateral Anterior Cerebral Artery Occlusion Resulting in Akinetic Mutism: A Case Report." *Acta Neurologica* 11: 252–58.

Jennett, B. 1991. "Vegetative State: Causes, Management, Ethical Dilemma." *Current Anaesthesiology and Critical Care* 2: 57–62.

Kinney, H. C., J. Korein, A. Panigrahy, P. Dikkes, and R. Goode. 1994. "Neuropathological Findings in the Brain of Karen Ann Quinlan: The Role of the Thalamus in the Persistent Vegetative State." *New England Journal of Medicine* 330: 1469–75.

Kinomura, S., J. Larsson, B. Gulyas, and P. E. Roland. 1996. "Activation by Attention of the Human Reticular System and Thalamic Intralaminar Nuclei." *Science* 271: 512–15.

Kuwert, T. 1991. "Regional Cerebral Glucose Consumption Measured by PET in Patients with Unilateral Thalamic Infarction." *Cerebrovascular Disease* 1: 332–36.

Libet, B. 1978. "Neuronal vs. Subjective Timing for a Conscious Sensory Experience." In P. A. Buser and A. Rougeul-Buser, eds., *Cerebral Correlates of Conscious Experience*. Amsterdam: Elsevier/North-Holland, 69–82.

Lisovoski, F., P. Koskas, and T. Dubard. 1993. "Left Tuberothalamic Artery Territory Infarction: Neuropsychological and MRI Features." *European Neurology* 33: 181–84.

Milhaud, D., G. Berrardin, and P. M. Roger. 1993. "Traitement d'un état de mutisme akinetique par la bromocriptine." *Presse Medicale* (Paris) 22: 688–91.

Multi-Society Task Force on PVS. 1994. "Medical Aspects of the Persistent Vegetative State." *New England Journal of Medicine* 330: 1499–1508.

Munk, M. H. J. 1996. "Role of Reticular Activation in the Modulation of Intracortical Synchronization." *Science* 272: 271–74.

Nieuwenhuys, R. J., J. Voogd, and S. van Huijzen. 1991. *Das Zentralnervensystem des Menschen*. Berlin: Springer-Verlag.

Plum, F., and J. B. Posner. 1980. *The Diagnosis of Stupor and Coma*. Philadelphia: F. A. Davis.

Posner, M. I., and M. E. Raichle. 1994. *Images of Mind*. New York: Scientific American Library.

Sandson, T. A. 1991. "Frontal Lobe Dysfunction Following Infarction of the Left-Sided Medial Thalamus." *Archives of Neurology* 48: 1300–1303.

Schoenle, P. W. "Neuropragmatics." *Neurologie und Rehabilitation* (in press).

Shimamura, A. P. 1995. "Memory and Frontal Lobe Functioning." In M. Gazzaniga, ed., *The Cognitive Neurosciences*. Cambridge, Mass.: MIT Press, 803–13.

Szelies, B. 1991. "Widespread Functional Effects of Discrete Thalamic Infarction." *Archives of Neurology* 48: 179–82.

Zschocke, S. 1995. *Klinische Elektroenzephalographie*. Berlin: Springer-Verlag.

Index

Achacoso, T., 204
Ackerman, R., 79n4
action potentials, model of, 200
AD. *See* Alzheimer's disease
Adams, P., 159
Adrian, E., 13
agnosia, 100, 273. *See also* aphasia
AI. *See* artificial intelligence
akinetic mutism, 270–72
Alberts, B., 225n3
algorithm, 58, 77, 157, 171n1
Allman, J., 108n19
Alsted, J., 23
Alzheimer's disease (AD), animal model of,
   218–21, 225n5; definition of, 218; genetic causes of, 218–19
α-amino-3-hydroxyl-5-methyl-4-isoxa-
   zolepropionic acid (AMPA) receptor:
   and LTP, 135n4; role of in spatial memory, 116, 117. *See also* spatial memory
Andersen, R., 160–61, 162
Anderson, R., 285n54
animal models, 31, 38, 41, 231–45;
   differences between, 31–38; extrapolation from, 210–24, 236; hierarchy of
   mechanisms in, 240–45; and theory
   construction, 31–41. *See also Aplysia*

*californica; Caenorhabditis elegans;
   Drosophila melanogaster;* model organisms; *Mus musculus*
Ankeny, R., 202, 205
antireductionism, and simulation, 138
aphasia, 12. *See also* agnosia
Aplysia californica (sea hare): and memory
   formation, 15; as model system, 200,
   236, 239. *See also* animal models; model
   organisms
architecture of mind, 8, 23–24
Aristotle, 4, 7, 24n1
ars memorativa, 8; and architecture of
   mind 22–24; and localization, 23–24;
   and fragmentation of mental faculties,
   23–24
artificial intelligence (AI), 1, 14. *See also*
   computation, neuronal; neuronal networks; simulation models
artificial neural networks, 148–49
artificial systems, compared with natural
   systems, 149–51
Ascaris suum, 203, 208
a-semantic (algorithm-semantic) interpretation, 157–61
associative mind, concept of, 7–9, 16, 17,
   18, 24

289